D1294385

ECONOMIC DEVELOPMENT & GIS

J.M. POGODZINSKI | RICHARD M. KOS, AICP

Esri Press
REDLANDS|CALIFORNIA

Esri Press, 380 New York Street, Redlands, California 92373-8100
Copyright © 2013 Esri
All rights reserved. First edition 2013
17 16 15 14 13   1 2 3 4 5 6 7 8 9 10

Printed in the United States of America

Library of Congress Cataloging-in-Publication Data
Pogodzinski, J. M.
Economic development and GIS / J.M. Pogodzinski, Richard M. Kos.
     p. cm.
     Includes bibliographical references and index.
     ISBN 978-1-58948-218-0 (pbk. : alk. paper)—ISBN 978-1-58948-332-3 (electronic)
     1. Industrial location—Planning—Geographic information systems. 2. Regional planning—Geographic information systems.
3. Business planning—Geographic information systems. 4. Economic development. 5. Urban economics. I. Kos, Richard M., 1967– II. Title.
     HD58.P63 2013
     338.900285--dc23   2012020286

The information contained in this document is the exclusive property of Esri unless otherwise noted. This work is protected under United States copyright law and the copyright laws of the given countries of origin and applicable international laws, treaties, and/or conventions. No part of this work may be reproduced or transmitted in any form or by any means, electronic or mechanical, including photocopying or recording, or by any information storage or retrieval system, except as expressly permitted in writing by Esri. All requests should be sent to Attention: Contracts and Legal Services Manager, Esri, 380 New York Street, Redlands, California 92373-8100, USA.

The information contained in this document is subject to change without notice.

U.S. Government Restricted/Limited Rights: Any software, documentation, and/or data delivered hereunder is subject to the terms of the License Agreement. In no event shall the U.S. Government acquire greater than restricted/limited rights. At a minimum, use, duplication, or disclosure by the U.S. Government is subject to restrictions as set forth in FAR §52.227-14 Alternates I, II, and III (JUN 1987); FAR §52.227-19 (JUN 1987) and/or FAR §12.211/12.212 (Commercial Technical Data/Computer Software); and DFARS §252.227-7015 (NOV 1995) (Technical Data) and/or DFARS §227.7202 (Computer Software), as applicable. Contractor/Manufacturer is Esri, 380 New York Street, Redlands, California 92373-8100, USA.

@esri.com, 3D Analyst, ACORN, Address Coder, ADF, AML, ArcAtlas, ArcCAD, ArcCatalog, ArcCOGO, ArcData, ArcDoc, ArcEdit, ArcEditor, ArcEurope, ArcExplorer, ArcExpress, ArcGIS, ArcGlobe, ArcGrid, ArcIMS, ARC/INFO, ArcInfo, ArcInfo Librarian, ArcLessons, ArcLocation, ArcLogistics, ArcMap, ArcNetwork, *ArcNews*, ArcObjects, ArcOpen, ArcPad, ArcPlot, ArcPress, ArcPy, ArcReader, ArcScan, ArcScene, ArcSchool, ArcScripts, ArcSDE, ArcSdl, ArcSketch, ArcStorm, ArcSurvey, ArcTIN, ArcToolbox, ArcTools, ArcUSA, *ArcUser*, ArcView, ArcVoyager, *ArcWatch*, ArcWeb, ArcWorld, ArcXML, Atlas GIS, AtlasWare, Avenue, BAO, Business Analyst, Business Analyst Online, BusinessMAP, CommunityInfo, Database Integrator, DBI Kit, EDN, Esri, Esri—Team GIS, Esri—*The GIS Company*, Esri—The GIS People, Esri—The GIS Software Leader, FormEdit, GeoCollector, Geographic Design System, Geography Matters, Geography Network, GIS by Esri, GIS Day, GIS for Everyone, GISData Server, JTX, MapIt, Maplex, MapObjects, MapStudio, ModelBuilder, MOLE, MPS—Atlas, PLTS, Rent-a-Tech, SDE, SML, Sourcebook·America, Spatial Database Engine, StreetMap, Tapestry, the ARC/INFO logo, the ArcGIS logo, the ArcGIS Explorer logo, the ArcPad logo,the ArcGlobe logo, the Esri globe logo, the Esri Press logo, the GIS Day logo, the MapIt logo, The Geographic Advantage, The Geographic Approach, The World's Leading Desktop GIS, *Water Writes*, www.arcgis.com, www.esri.com, www.geographynetwork.com, www.gis.com, www.gisday.com, and Your Personal Geographic Information System are trademarks, registered trademarks, or service marks of Esri in the United States, the European Community, or certain other jurisdictions. Other companies and products mentioned herein are trademarks or registered trademarks of their respective trademark owners.

Ask for Esri Press titles at your local bookstore or order by calling 800-447-9778, or shop online at www.esri.com/esripress. Outside the United States, contact your local Esri distributor or shop online at www.eurospanbookstore.com/esri.

Esri Press titles are distributed to the trade by the following:

In North America:

Ingram Publisher Services
Toll-free telephone: 800-648-3104
Toll-free fax: 800-838-1149
E-mail: customerservice@ingrampublisherservices.com

In the United Kingdom, Europe, Middle East and Africa, Asia, and Australia
Eurospan Group
3 Henrietta Street
London WC2E 8LU
United Kingdom
Telephone: 44(0) 1767 604972
Fax: 44(0) 1767 601640
E-mail: eurospan@turpin-distribution.com

ACC LIBRARY SERVICES AUSTIN, TX

Contents

Foreword

Spatial statistics and econometrics have been highly developed over the past half century, yet these techniques have only recently proved routinely useful for economic analysis and applied research. The relatively slow pace of this advance arises in part from inherent difficulties in the diffusion of progress in applied statistical methods. In some part, however, this slow development arises from the difficulties in computation and software development.

This volume, *Economic Development and GIS*, explains the use of geographical information techniques by combining clear applications of policy problems with direct approaches, using specific computational techniques. The authors are careful to identify the problems that can be addressed more directly and powerfully by using GIS (geographic information systems) than by using more primitive techniques. In this way, the book provides a vivid illustration of the value of geographical information techniques in economic development, in which the analysis of aggregate spatial relationships is so important.

Many of the advantages in applying GIS techniques do not arise from the geographic aspects of the problems and solutions. Rather, the advantages often arise from the clear organization of data implied by GIS techniques and the comprehensive view of problems and the data necessary to solve those problems. GIS provides a discipline that helps users organize and focus on the data required for efficient decision making. With GIS, users direct their collaborative efforts to solve problems across agencies, disciplines, and geographic regions. Throughout the discussion, the authors highlight the tension between the *problems* to be solved and the *methods* to be applied, building upon simple and then increasingly complex examples of GIS in action.

The first chapters review the links between tools and methods and introduce ArcGIS software and other principal software tools used for analysis. These tools from Esri are available online and in various desktop versions. Later chapters discuss specific applications of GIS to the analysis of economic development. Examples include selecting a site for a retail outlet, designating an enterprise zone, and understanding issues involved in balancing jobs and housing.

These chapters provide examples of the use of GIS techniques and software to address specific problems of broad concern in economic development. The book discusses data interoperability and the capacity of ArcGIS software to read and process non-native formats for data, which is helpful in assembling datasets from many different sources and programs. The book also features examples of using image (or raster) files in the analysis of spatial problems and highlights the analytical advantages of manipulating raster data.

The highlight of *Economic Development and GIS* is the seamless integration of chapters contributed by practitioners and development specialists to show why advanced techniques are used to address a host of problems in theory and practice. As a blend of new ideas and applications, the book is a treatise on the power of GIS to help organize problems for systematic analysis.

John M. Quigley
I. Donald Terner, Distinguished Professor
University of California
Berkeley

Acknowledgments

Thanks to Jennifer Chen, Julie Amato, Leif Schmit-Kallas, Jean Casey, Bette McDonnell, Eric J. Kos, Sonja Caldwell Kos, Diana Pancholi, Paul Hierling, Jeff Johnson, Travis Miller, Matthew Gardner, John Roach, Matthew Perry, and the excellent editors at Esri Press.

Introduction

This book demonstrates why GIS (geographic information systems) is an essential tool for modern economic development analysis; all such analyses now involve at least some GIS methods and tools. The book pursues this purpose full in the knowledge that economic development analysis has been undertaken successfully in the past by people without any knowledge or use of GIS.

Economic development is inherently a team effort, requiring collaboration across professions, departments, divisions, and communities. This book illustrates how GIS promotes and increases teamwork among those with diverse expertise. Knowing just a few basic concepts and applications of GIS contributes substantially to a project's success. Yet even experienced users may not fully understand the capabilities that GIS offers as a tool for confronting economic development issues at the local, regional, and state levels—and beyond. As numerous examples in these pages show, the task of analyzing these issues is easier and more effective with GIS than without it.

The book highlights the successful use of ArcGIS, a software suite from Esri that implements the concepts of GIS. ArcGIS software—the most widely used and powerful GIS program—makes easily available the features and tools of GIS that are most important for economic development analysis. Some of our discussion (for example, the discussion of Esri Business Analyst in chapter 2) is specific to ArcGIS software.

What makes GIS essential?

Economic development analysis was performed for decades without GIS. What has happened in recent years to make GIS the essential tool for economic development analysis? Briefly, the answer is that more data are available, and more sophisticated techniques are available to analyze the data, so more is expected from economic development officials.

GIS plays a key role in economic development analysis because it

- accounts for the spatial aspects of economic development;
- promotes teamwork, data-driven decision making, and a more holistic and inclusive view of problems;
- helps organize data and workflows;
- covers larger areas and incorporates more information than traditional economic analysis;
- incorporates vector and raster data (imagery) and data arising from diverse sources, such as GPS loggers;
- requires only a few GIS tools to undertake a sophisticated economic development analysis.

As an example of the last point, chapter 1 uses only *seven* tools of ArcGIS to complete an economic development analysis—one of several analyses using these tools that could be carried out. These seven tools are easy to learn and provide great insight into the capabilities of GIS.

Every chapter of the book addresses at least one of the key points listed above.

The book juxtaposes two themes: problems and methods. People tend to think of problems in terms of the methods they have at hand to solve them. Often a problem seems too abstract or too difficult if there is no ready method for solving it. At the same time, few people really take an interest in the contents of a toolbox—screwdrivers, wrenches, wire cutters, and so on. True, these tools are useful, and *the initiated* understand their value. In this spirit, this book pulls the tools of GIS out of the box to explain their use—and the results of their use—for analyzing and solving problems in economic development.

How the book is organized
The book is divided into three parts, followed by an afterword.

Part I broadly reviews economic development analysis and GIS tools and methods used to carry out the analysis. This part introduces economic development and "the Geographic Approach," which forms the basis for GIS analysis. Chapter 1 provides an overall framework for understanding the role of GIS in economic development analysis. Chapter 2 discusses Business Analyst (available online and in a desktop version), which makes data and tools for economic development analysis immediately available. Chapter 3 covers best practices for leveraging the capabilities of GIS in project organization, data management, and group collaboration.

Part II discusses applications of GIS to economic development analysis. The three chapters in this part stand alone but also broaden the discussion in the first part by introducing a new aspect of economic development analysis, using new GIS tools, or doing both. Chapter 4 discusses site selection—a core problem in economic development analysis—in terms of identifying a suitable location for a hypothetical retail store. Chapter 5 describes how GIS can, in a straightforward manner, identify an area for an enterprise zone that has to meet multiple criteria. The tools used to identify the zone can also be used to undertake "what if?" analysis if the criteria change. Chapter 6 discusses the integration of GIS with GPS technology and data in transportation issues, including jobs-housing balance, transit-oriented development, and commute time.

Part III explores the tools of GIS more fully. The final three chapters take the reader a step or two beyond the basic applications. Chapter 7 discusses geocoding—putting digital pins on a map based on addresses or other location-based data—and how geocoding opens a door to a wide range of analyses that are important to economic development. Chapter 8 examines how statistical analysis can be undertaken in GIS. Chapter 9 explains that raster data (like aerial and satellite imagery) can serve as more than a detailed backdrop to maps; it can provide analytical possibilities in economic development.

The book concludes with an afterword that sums up the relationship between GIS and economic development, looks ahead to other issues in economic development (sustainability and regional interrelationships), and discusses how GIS could impact those issues.

The unique features in this book include, most notably, the following:

- Application of Esri Business Analyst Desktop (BA Desktop) and Esri Business Analyst Online (BAO) to economic development problems
- Discussion of statistical applications to economic development with GIS examples relevant to economic development
- Creation of data using GIS that could not have been created in another way
- Application of ArcGIS Spatial Analyst to economic development with focus on the analytical possibilities afforded by the unique structure of raster files
- Application of ArcGIS Data Interoperability for Desktop to economic development, which makes interagency regional cooperation involving GIS data easier

The great wealth of topics includes core concepts and proven applications in economic development—as well as the proper context for approaching many economic development problems. Our hope is that by the end of this book, readers will gain a deeper appreciation of the value and use of GIS in economic development for assessing, sustaining, and improving the quality of life.

Part I
Economic development, the Geographic Approach, and GIS

1

Introduction to economic development and GIS

Objectives

- Outline the main economic theories and concepts used in economic development analysis
- Illustrate economic concepts of clustering, agglomeration economies, spillovers, scale economies, and related concepts within the context of GIS
- Demonstrate the application of GIS to workforce development

An economic development problem

Economic development officials in Silicon Valley face the problem of retaining biotechnology firms in the area and attracting more such firms. Most people associate the Silicon Valley,[1] just south of San Francisco, with microchip design and manufacturing. Less well-known, perhaps, is that the area is also a world center of the biotechnology industry.

Although already a leader in biotech, Silicon Valley faces several challenges in expanding the industry. Perhaps the most important constraint is the limited availability of appropriately trained mid-level "white lab coat" technicians to work in biotechnology firms. (A similar constraint is faced by biotech clusters almost everywhere in the world.) **Workforce development**—investment in training and education to meet industry needs—is a key economic development policy for retaining and expanding the biotechnology industry.

This chapter explains why GIS is essential in economic development analysis and examines the spatial (location-based) aspects of economic development and the way these spatial aspects are represented in GIS and reflected in GIS-supported economic development analyses. This chapter will present an example to show how a workforce development strategy for biotechnology in Silicon Valley can be analyzed with the use of GIS methods and tools.

Biotechnology is a desirable industry. It is not a "smokestack" industry in that it involves relatively little emissions pollution. It is a growing industry in contrast to the declining heavy manufacturing sector. As a global industry, biotechnology contributes strongly to a region's exports. Furthermore, biotechnology jobs offer relatively high pay, even for mid-level workers.

There are several reasons grounded in economic development theory for the emergence of a biotechnology cluster in Silicon Valley. As shown in the example of workforce development policy described later in this chapter, the economic development theories and concepts that explain the location of industries have a strong spatial component that makes them particularly suited for representation and analysis using GIS.

Some questions and answers about economic development and GIS

Public officials and voters are often faced with a bewildering array of choices about allocating limited public resources. People want a higher quality of life, education for their children, and employment for themselves and other community members. These goals, all involving public resources, are promoted under the broad category of "economic development."

What is economic development?

The term "economic development" suggests different things to different people. At its broadest level, economic development is associated with an increase in real income per person, usually measured as **gross domestic product (GDP)** per person.[2]

More narrowly, the term suggests increased economic activity (sometimes loosely defined) within a relatively small geographic area, such as a redevelopment area or a state enterprise zone. The increased economic activity may be measured by jobs created, income generated, increased sales, or increased local revenues from particular taxes.

Fostering economic development involves interrelated choices by a variety of actors. Governmental policy actions, decisions by individuals, and industry location decisions must jibe if community members are to acquire skills needed by an emerging industry and receive stable and high-paying employment. Firms tend to locate near sources of available and appropriately skilled labor, and the presence of firms requiring a certain set of skills gives incentives to individuals to acquire those skills. Public policies support colleges and universities, which are the main vehicles for acquiring work skills.

What is GIS?

GIS software is a collection of programs that render (display) spatial data and support analysis of spatial (location-based) relationships, including sophisticated statistical analyses. One way to think about GIS programs is that they are like spreadsheets for maps. Spreadsheets are widely used in economic development analysis to record data, analyze data, make projections, undertake "what if?" analysis, prepare reports and graphics that support policy recommendations, document methods, and share both data and methods with collaborators and the general public.[3]

GIS software does all of these things but in a way that keeps track of location. GIS also includes built-in tools and functions for working with, representing, and analyzing spatial data. As with a spreadsheet, GIS can (and we argue *should*) be used to manage both spatial and nonspatial data. GIS software can also promote collaboration within teams that include non-GIS specialists (see chapter 3). GIS also provides ways to share data, methods, and results with the general public.[4]

Why is GIS essential to economic development analysis?

Spatial analysis is important in formulating economic development policies. Many of the concepts employed in economic development analysis have inherent spatial components. GIS is often the best, and sometimes only, way to handle these spatial elements. GIS is an efficient tool for economic development analysis for at least six reasons:

1. GIS can be used to easily link graphics on maps (such as the locations of biotech firms) to tabular records in databases (such as the company name and activity levels at each biotech firm).

2. GIS can be used to link features based solely on spatial location. This means that GIS can easily create datasets that would otherwise be difficult or impossible to create, as shown in examples later in this chapter and more fully in a discussion about GIS-created datasets in chapter 8. These data can also be exported from GIS in a variety of formats for use in other applications, for example, Excel or statistical software packages, such as SPSS, Stata, or R.

3. GIS can be used to track quantitative data over time, such as changes in land values, and quickly generate maps and graphs that "tell the story."
4. Analyses of spatial relationships involving concepts of distance, proximity, areas of overlap, and areas of concentration are relatively easy using GIS. Some advanced tools in ArcGIS software employ sophisticated spatial statistical methods.
5. GIS makes it easy to perform "what if?" analysis of alternative economic development projects and policies.
6. GIS maps can be generated in several formats, including interactive maps that can be posted on a website.

   This book will explain the applications of GIS to economic development problems that illustrate each of the capabilities just listed.

How does economic development analysis using GIS compare to economic development analysis without GIS?

As noted in the introduction, economic development analysis has been done for many years—and continues to be done today—without GIS. The next section on site selection identifies some characteristic problems or issues confronted by economic development officials and compares how those problems or issues have been handled with and without GIS.

Site selection

Site selection is the problem of finding the right location for a retail store, manufacturing firm, public facility, golf course, apartment building, and so forth. In one way or another, most economic development issues are related to site selection. Local officials are often called upon to identify specific parcels that have a particular combination of characteristics. Economic development officials perform site-suitability analyses. Often, multiple criteria must be satisfied. Site selection can be done without GIS—in fact, without any software or electronic database. Officials simply know the ground they are responsible for and identify a few sites that are suitable for a project.

However, there are inherent limits to "knowing the ground" in this way. Local development officials will often be called upon to compare sites in the local area with sites far away, even across the country. Local knowledge will not be enough for this purpose. Furthermore, more data are required as the criteria for suitability increase in number and complexity and as the size of economic development areas increases. Some of these data may be available only in digital form and require calculations of proximity that cannot readily be done by hand.

Moreover, the suitability of a site may need to be formally documented, for example, in a grant application. This requires access to a large body of timely, accurate, and sourced data. Manipulating these data requires analytic tools, and efficiently producing presentation-ready reports requires software that features report-writing capabilities. Examples of graphics generated by ArcGIS software will be shown later in this chapter, and ArcGIS's report-writing capabilities are more fully explored in chapters 2 and 4.

Site selection also relies on understanding the basic geography of an area, including its rivers, hills, earthquake fault lines, and so forth. At the municipal level, economic development involves land management in terms of a future-oriented general plan and current zoning regulations. Land-use ordinances dictate how land can be developed.

GIS allows a community's geographic features and land-use designations to be displayed and manipulated readily. In identifying or suggesting suitable sites for a particular development activity, the location of complementary activities or the location of competitors may be crucial to the viability of the proposed sites. GIS makes it possible (by using selection tools) to easily display a variety of characteristics and also to select a combination of desired characteristics (such as elevation, soil type, slope, and so on).

Zoning

Determining the appropriate zoning for a parcel or larger area depends on accounting for the proximity of a proposed activity to other activities or individuals. GIS allows for the easy computation of a variety of distance measures determined by the analyst (for example, straight-line distance or drive-time distance). GIS tools, such as **geocoding** (the creation of digital "pin maps" showing discrete locations), allow the display and analysis of the spatial distribution of firms. The spatial distribution of population affects the choice of location by firms in obvious ways (retail firms want to be close to their customers) and in more subtle ways (the concentration of population may affect firms' costs). Economists use the term "economies of urbanization" to describe the effect on costs of the concentration of population; this concept is discussed in greater detail later in this chapter. GIS tools allow users to identify and display the distribution of households with specific characteristics and determine their distances from various geographic features. Zoning conversions or variances are controversial because they affect property values and the quality of life in the immediate area. Partly for this reason, GIS methods are essential to visualize the impact of changes in zoning.

Externalities are spillovers from one land use onto adjacent users. Many production processes involve emissions, including noise, smoke, and odors. The rationale of land-use zoning is to segregate incompatible uses to minimize negative or undesirable spillovers of this sort. Limiting undesirable spillovers may be especially important if the generators of the undesirable spillovers are near sensitive receptors such as schools or facilities for seniors. One example of an analysis of spillovers is a GIS-based study of airflow patterns for particulate matter. The airflow patterns can be represented using GIS.

Fiscal impacts

Fiscal impacts are often spatial. For example, the impact (positive or negative) of a transportation hub on property values diminishes with distance from the hub. GIS can be used to model and visualize these impacts.

Fiscal impacts depend in large part on the transportation infrastructure. The movement of firms, people, and products is costly and is reflected in changes in the spatial distribution of firms and in altered demographics of neighborhoods. Cost of travel is affected by road capacity, transit infrastructure, density of population, and social and political characteristics. All these factors can be represented in a GIS framework. The significance of transportation infrastructure to firm location is illustrated in the example at the end of this chapter.

Fiscal impacts are rarely confined to just one political jurisdiction. Jurisdictional boundaries of governments (counties, community college districts, school districts, and so forth) typically overlap, and the overlapping (and often competing) political jurisdictions exert decision-making power over important variables that affect the viability of economic development projects. GIS tools can be used to examine the relationships among the overlapping political jurisdictions and the characteristics of firms and households across these areas. Chapter 8 has an example that shows the significance of jurisdictional boundaries.

Special zones

GIS can be used to determine the boundaries of an incentive area (like a redevelopment area, state enterprise zone, or brownfield), which may determine its fiscal viability. The spatial distribution of population and the ownership pattern of the land determine which individuals or socioeconomic groups bear the costs and which reap the benefits of an economic development project or policy. Population characteristics (such as income distribution) and ownership patterns (such as tenure status—renter or owner-occupied housing) are easily represented using GIS, facilitating the analysis of cost and benefit impacts.

Changing the boundaries of an incentive area may substantially affect the costs and benefits of the activity. The appropriate redevelopment area depends on several factors, such as the spatial distribution of existing firms and the characteristics of households in or near the redevelopment area. GIS can facilitate "what if" analysis, such as changing the configuration of the area while accounting for factors like the characteristics of the population affected by the adjustment.

Where can data be found to use in GIS analysis in economic development?

Datasets used in GIS analysis are often compiled from a variety of sources and include data of different types. Much of the data used in GIS analysis are available from public sources, including the US Census Bureau (see the sidebar, "Know the data," in chapter 5). All of the data used in the biotechnology workforce development example in this chapter come from public sources, except for the locations of biotech firms. Esri Business Analyst (a specialized application built on the core ArcGIS platform and discussed in chapter 2) comes bundled with data, including detailed business and forecast data for use in economic analysis.

The types of data used in GIS include **vector data** (map data that illustrate points, lines, and polygon shapes, such as customer locations, transportation routes, and jurisdictional boundaries, respectively) and **raster data** (pixel-based data, such as aerial or satellite images). This book explores various data types encountered in GIS applications. The workforce development policy analysis at the end of this chapter uses all types of vector data (points, lines, and polygons) along with **tabular (table) data** about demographic characteristics of the population.

What methods of GIS are most useful in economic development analysis?
The seven tools

ArcGIS includes hundreds of tools. However, important analyses of economic development projects and policies can be undertaken using only a handful of these tools. The workforce development application presented in this chapter was completed using only seven of these tools. Indeed, *different* economic development analyses can be undertaken by using the *same* seven tools in different ways or in different combinations. These tools are very important, and the range of possible analyses that can be generated using these tools in different ways and combinations is intended to inspire readers to learn still more tools to undertake even more far-reaching analyses. (Later chapters give several illustrations of such extensions using additional GIS tools.)

The tools used in the workforce development application were the following:

- Table Join—the process of joining data from a table to a map layer, based on a common field (variable)
- Spatial Join—the process of joining data from different map layers, based on location only (without necessarily having a common field)
- Geocoding—the process of creating digital "pins," based on address information
- Select by Attributes—the process of identifying subsets of a table or map layer, based on a characteristic the user specifies
- Symbolizing quantitative data—the process of visualizing quantitative data by using various shadings, colors, symbols, or sizes to reflect different quantities of a variable
- Buffer—the process of creating areas of a given distance around objects identified in map layers, such as rings of a given radius around a particular site
- Intersect—the process of finding the common part of two map layers, analogous to the intersection of two sets in mathematics

Spatial statistics tools

ArcGIS contains several statistical tools that have important applications in economic development. Chapter 8 is devoted to an application of some of these tools to the same workforce development issue discussed in this chapter. Chapter 8 focuses on tools that are used to analyze patterns and clusters, especially **hot spot analysis**. Chapter 8 also deals with modeling economic relationships in a spatial context with **regression analysis**.

Using raster data in economic development analysis

Rasters, as mentioned above, are pixel-based data, including aerial and satellite images. Rasters can enhance the look of a map, but they have a use beyond making pretty maps. Chapter 9 illustrates the powerful analytic tools that can be employed with raster data in economic development.

Sources of economic development

Listed here are a few of the most salient factors that account for economic development.[5] These factors are key building blocks of the economic theories of economic development.

Increases in labor productivity

When labor becomes more productive in the sense that an hour of a worker's time results in more value produced, there has been an increase in labor productivity. Generally, in the United States, increases in real gross domestic product (GDP) result from increases in **labor productivity.** There are three main sources for such an increase in productivity. First, each worker may use more or better capital equipment (machines) that allows an hour's work to produce more output or more valuable output. So the ability economy-wide to accumulate more capital for workers to use is one of the key elements in macroeconomic growth. A second source of increased labor productivity touches on the topic discussed in detail in the example at the end of this chapter. Workers may produce more value in an hour's time because they have more skills—more **human capital**, as opposed to machine capital. Workforce development is the process of promoting the acquisition of skills by workers. Third, technological change—new ways of producing goods and services—is a source of increased labor productivity. The term for technologically driven increases in productivity is multifactor productivity. It is difficult to identify the separate contributions to growth of technology, physical capital, and human capital. To employ a new technology, one typically has to have new machines, which embody the technology, and newly trained workers, who are adept at the new technology.

Animal spirits

"Animal spirits" was a term coined by John Maynard Keynes, the father of macroeconomics,[6] to describe the motivation and risk-taking behavior of investors. More recently, the phrase has inspired further research based on the newly emerging discipline of behavioral economics.[7] The term alludes to the essential role of risk-taking in economic investment, and more broadly, to the role of psychological factors (the mass psychology of the market) in explaining booms and busts in economic activity.

Social networks

Sociologists, geographers, psychologists, and others have been delving into models of relationships—professional, social, and ethnic—that support innovation and entrepreneurship. There is an emerging literature[8] on the roles such social networks play in the development of industries such as biotechnology and clean technology.

Economic gardening

In this same vein, over the last dozen years, a movement espousing an "economic gardening" approach to economic development has arisen. The economic gardening model has much in common with the "animal spirits" focus of recent behavioral economics and with social network analysis. The focus of economic gardening is on entrepreneurship and, in particular, on the role of local governments and organizations to promote entrepreneurial endeavors.

Fundamental economic concepts in economic development

Before continuing on with the analysis of workforce development policy in the biotechnology industry of Silicon Valley, the chapter will review a few key economic concepts and their spatial implications, which will be used in that analysis and at various points throughout the book.

Transportation cost and market area analysis

Space matters in economics because it is costly to overcome distance, and proximity may have benefits or costs. The costliness of distance is affected by technology. Old cities were limited to the distance people could travel by walking or using animal power. Later-developing cities, which expanded after the advent of the automobile, became much more spread-out as the cost of transportation fell dramatically and the increased speed and reliability of transportation made movement of people and things easier. Public infrastructure development, like a highway, may reduce transportation costs along a certain route and thereby encourage firms and households to locate along that route. This is a factor in the location of biotechnology firms. (This is called **transit-oriented development** and is discussed more fully in chapter 6.)

GIS can be used to represent the road and highway network, including capacity indicators. Road capacity and characteristics determine transportation cost along various corridors. GIS can be used to determine driving distances, and the Esri program ArcLogistics can analyze alternative routes. Transportation costs are one of the central elements in determining **market areas** for retail firms.

Scale economies

Scale economies refers to the decrease in *average* cost of production (per unit cost) as the scale (or number of units produced) increases. For example, the average cost of publishing a book decreases as the print run—how many copies are produced—increases. This is because in book publishing there are many fixed costs, and these fixed costs are a large part of the total cost of the book publishing project. The costs of writing, editing, and preparing a manuscript for publication are fixed: these costs must be paid, regardless of the number of copies of the book that are printed. Average cost is the total cost of the project divided by the number of units produced. The greater the print run, the larger the number used to divide the fixed costs. It is assumed that the variable costs, for example, the cost of paper and ink, are a small part of the total cost of the project. Scale economies mean that *size does matter.*

The significance of scale economies for economic development analysis cannot be overstated, whether viewed from a macro-scale of a state or nation or from a micro-scale of a local economic development project. The macro-effects of scale economies are evident everywhere: cities exist because of scale economies. Economic development is concentrated in some regions partly because of scale economies.

Regions, and even municipalities, may also be viewed as trading entities, which export and import different goods and services. Scale economies constitute a basis for trade. The traditional basis for trade is comparative advantage—differences in opportunity costs across regions. However, more recent theory in international trade focuses on the role of scale economies and product differentiation.

On the local level, the presence of economies of scale (lower average cost associated with size) argues for larger rather than smaller projects because of lower average costs associated with larger projects. For example, large housing development projects are likely to have lower average costs than in-fill projects.

Agglomeration economies

Agglomeration economies are a special kind of external scale economy—the scale is associated with the concentration (number or density) of people or firms in an area. If agglomeration economies are present, the average cost of production of firms decreases as the level of agglomeration (the number or density) of firms or people increases. There are two types of agglomeration economies: economies of localization and economies of urbanization.[10]

Economies of localization refer to the decrease in average cost of firms in a *particular* industry associated with the concentration of firms in the *same* industry nearby. Economies of localization account for industry clusters, like the clustering of semiconductor design and manufacturing firms in Silicon Valley and the clustering of biotechnology firms in the San Francisco Bay Area. Economies of localization refer to a specific industry and therefore are driven by factors particular to that industry.

"Industry" can be defined more broadly or more narrowly, so our appreciation of the extent of localization economies of scale may be affected by the definition of industry that is used. (See the discussion of North American Industry Classification System [NAICS] Codes in this chapter). Economies of localization in one industry may be associated with economies of localization in another industry, if the industries are related in the sense that the underlying causes of reduced average cost in the first industry spill over onto the other industry.[11] Not only is Silicon Valley an agglomeration of microprocessor design and manufacturing firms, but it is also an agglomeration of biotechnology and green technology firms. For example, one can speak broadly of the "restaurant industry" or more narrowly of the "fast-food industry." Practically, industry statistics can be examined at various levels of disaggregation corresponding to digit levels of NAICS Codes (for example, two-digit level [very broad] or six-digit level [very narrow]).

NAICS Codes

NAICS Codes are the US government's official numerical designation for classifying firms into industry groups. This system is the successor to the US government's Standard Industry Classification (SIC) Codes that were used until 1997. Under either classification system, "industry" could be defined very broadly at the "two-digit" level, more narrowly at the "four-digit" level, or even more narrowly at the "six-digit level."

For example, most broadly, at the two-digit level, biotechnology is contained within NAICS Code 54, which encompasses "Professional, Scientific, and Technical Services" that includes legal and accounting services. More narrowly, at the four-digit level, biotechnology is contained within NAICS Code 5417, for "Scientific, Research, and Development Services." The most refined industry classification available for biotechnology is the six-digit NAICS Code 541711, for "Biotechnology Research and Development."

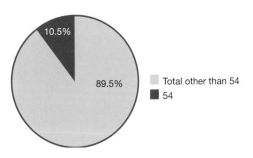

Figure 1S-1. NAICS Code 54 shown as a proportion of the total economy in terms of payroll in 2008. Courtesy of US Census.

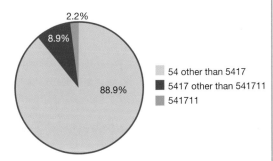

Figure 1S-2. Showing two-digit, four-digit, and six-digit levels of NAICS Code 54 in terms of payroll in 2008. Courtesy of US Census.

Firm clustering can be represented and analyzed using GIS. Our example at the end of this chapter and the more sophisticated statistical analysis in chapter 8 illustrate and analyze clustering of biotechnology firms.

The effects of economies of localization and the factors that account for localization economies, like the availability of specialized workers (the spatial extent of the labor pool), can be represented using GIS. The concentrations of firms of a particular type, the concentration of people with particular characteristics, and areas of different sizes can all be displayed and analyzed using GIS. Our biotechnology workforce development example at the end of this chapter illustrates these capabilities of GIS.

Economies of urbanization refer to the decrease in average cost associated with the concentration of people or firms—or both generally—not with reference to any specific industry. Some activities may exhibit lower average cost simply because of the concentration of population. Public transit is a good example. Usually, the average cost of providing public transit services is lower in areas where the population density is higher along the public transit route. Ridership will be greater if the route passes through a dense urban area rather than a sparsely populated suburban or exurban area. Thus, the average cost per rider per mile will be lower. The decline in average cost is associated solely with the concentration of people, not with the concentration of firms in a particular industry.

Externalities or spillovers

Externalities (or spillovers) are the positive or negative effects one firm or household has upon another firm or household that are not fully compensated for (in the case of a negative externality) or paid for (in the case of a positive externality) by the recipient. For example, regarding the flu vaccine, there are benefits from contacting people who have had the flu vaccine (compared to those who haven't) because their chances of getting the flu and spreading it are lower. In other words, the person getting a flu vaccine provides an external benefit (or positive externality) to those who did not participate in (or whose interests were not necessarily accounted for) in the decision to get the vaccine.

Externalities—both positive and negative—are particularly important in economic development. Should funds to rehabilitate dilapidated housing be expended roughly equally across the city, or should they be concentrated in a few areas? Someone interested in maximizing the increase in property values (and perhaps the associated increase in property tax revenues) must account for the (positive) externality that an improvement in one house has on the value of neighboring houses. If the goal is to maximize the increase in house values for a given housing rehabilitation budget, expenditures should be concentrated in a few areas—the houses in the rehabilitated neighborhoods will exert positive externalities upon one another—rather than spread equally (in which case the positive externality effect would be more quickly dissipated). Concentrations of rehabilitated and dilapidated housing can be represented in GIS along with the distribution of housing prices.

"Knowledge spillovers,"[12] one of the most important kinds of externalities that account for clustering of some industries, have a spatial aspect. They suggest that some industries may be concentrated around universities.[13] Knowledge spillovers have been tracked and analyzed using GIS.[14] The example at the end of this chapter looks at the proximity of biotech firms to research universities.

Mobility, sorting, and transaction costs

Households and firms can move in response to changes in taxes, public spending on roads, crime, and a host of other factors. This potential for movement complicates the analysis of economic development projects and policies. As illustrated in this chapter, households tend to sort themselves into distinct classes, and the sorting can be visualized using GIS. Moreover, economic development projects and policies impact the location choices of households. GIS can be used to trace the effects on households' location choices over time. GIS provides an array of tools and methods to analyze the degree of concentration along several dimensions.

Biotechnology industry clustering and workforce development policies: A case study of economic development and GIS

GIS will be used to examine the clustering of biotech firms in the San Francisco Bay Area, based on the following factors:

- Funding of biology research
- Location and capacities of educational institutions with biology programs
- Demographic and economic factors, like age distribution and income
- Transportation facilities
- Housing costs

These relationships provide insight into workforce development and related economic development policies using the concepts and theories introduced above. Chapter 8 introduces methods to model the relationships among spatial variables.

The application in this chapter employs only the *seven* tools of ArcGIS mentioned above. Illustrating this discussion are the most parsimonious graphics, which display the use of the analytic tools rather than final, presentation-ready maps.

Data related to industry clustering and workforce development

Data employed in economic development analysis typically arise from different sources and differ in terms of geographic coverage, time period, level of aggregation, number of observations (records), and variables (fields) included. For example, the US Economic Census of 2002 gives the distribution of firms down to very refined industry classification[15] or ZIP Codes and provides the distribution of firms by revenues or sales for those firms in business for at least a year and a count of those in business for less than one year. The US Census County Business Patterns data gives counts of firms by employment level for each year by industry and ZIP Code, but does not provide information on firms' revenues, and does not identify the number of firms that are less than one year old.[16]

Esri Business Analyst Desktop (BA Desktop) and Esri Business Analyst Online (BAO) come with up-to-date, detailed, immediately useable data on business locations and characteristics for specific firms, as well as built-in tools for analysis. Business Analyst and its use in economic development are described in chapter 2 and elsewhere in the book.

The data used in this application come from several sources.[17] These sources provide information about degrees granted by discipline and institutions, the locations of institutions of higher education, locations and characteristics of biotechnology firms, and demographic and economic data about the population.

Public investment impacts biotech clustering in several ways. First, the clustering of biotech firms is partly fostered by public investment in research at university centers.[18] Second, the extent of clustering and the prospects for the growth of the industry depend on the supply of workers with specific skills. Third, public investment in transportation infrastructure affects industry location and therefore industry clustering. This section examines the funding of biology research, accessibility of worksites, and supply of graduates of public and private colleges and universities with degrees in biology. Taken together, these elements measure the capacity of educational institutions to provide skilled labor for biotech firms, the impact of research funding, and accessibility of work locations.

The economic significance of learning cannot be overstated. It is the foundation of individual, regional, and national advancement. In a speech in July 2009, President Obama emphasized the crucial role of community colleges in workforce development.[19] In announcing the American Graduation Initiative, the president said, "Community colleges are an essential part of our recovery in the present—and our prosperity in the future." The president noted that, ". . . jobs requiring at least an associate degree are projected to grow twice as fast as jobs requiring no college experience. We will not fill those jobs, or even keep those jobs here in America, without the training offered by community colleges." The American Graduation Initiative includes competitive grants to community colleges to pursue innovative programs, partnering with industry, creation of a research center focused on best practices, incentives for student completion of programs, and funding for infrastructure.

Simultaneously with the president's speech, the Council of Economic Advisors issued a report titled "Preparing the Workers of Today for the Jobs of the Future."[20] The report cites numerous studies indicating that the economy in 2016 will be more dynamic and more reliant on highly educated workers than the economy of today. "Worker flexibility is key, given the dynamic nature of the US labor market and ongoing technological change," the report states.[21] The report also examines the likely effects of the federal stimulus act, the American Recovery and Reinvestment Act (ARRA),[22] on workforce issues. Workforce development policy has been the subject of a vast literature. A recent salient contribution to that literature is the book *The Race Between Education and Technology,* by Claudia Goldin and Lawrence F. Katz.[23]

Workforce development can be viewed from several perspectives. Biotechnology firms decide whether to locate or expand operations in an area by comparing that area to alternative locations along several dimensions, including the availability of skilled workers. College and university administrators want to expand programs that are likely to attract enrollment and are therefore viewed as needed and successful. Students want to enter fields where stable and high-paying employment is likely, and they desire college and university programs that are located close to their homes or places of current employment. Transportation officials want to invest funding to best serve the public interest.

Biotechnology firms in the San Francisco Bay Area

In some ways the development of the biotechnology industry in Silicon Valley mimics that of microprocessor design and manufacturing. For both, the comparative advantages of Silicon Valley are based on human capital—the proximity of leading research universities; a creative class of trained workers; and a social, cultural, and economic infrastructure that supports entrepreneurship. The analysis will focus on biotech firms in the San Francisco Bay Area, which comprises nine counties: Alameda, Contra Costa, Marin, Napa, San Francisco, San Mateo, Santa Clara, Solano, and Sonoma.

Colleges and universities in the San Francisco Bay Area and biotechnology

Three major research universities are located in the San Francisco Bay Area: the University of California, San Francisco; the University of California, Berkeley; and Stanford University. Five campuses of the California State University (CSU) system and twenty-six public community colleges are also located in the region, along with several other private colleges and universities.

Figure 1-1 shows the nine Bay Area counties outlined in heavy black lines, with the locations of the three research universities mentioned above indicated by yellow dots. In a GIS program, each graphic feature displayed on a map (water feature, county boundary, and university location) is linked to a record in a **database** that provides detailed attribute information (contained in an **attribute table,** such as the name of the water feature; name and population of each county; and name, enrollment, grant funding level, and city of each university displayed. These map graphics are added as specific layers of information to ArcGIS. In this example, the county boundaries are one such layer (that is, a specific file that resides on your computer), the water features are another, and the universities are a third. ArcGIS automatically aligns these separate data layers to form the composite map seen in figure 1-1 because each dataset has spatial information that informs ArcGIS as to where each layer is located on the surface of the earth.

Each of the research universities receives substantial funding for research in medicine and biology; the totals range from about $500 million per year for the University of California at San Francisco to over $400 million per year for Stanford to just over $150 million per year for the University of California at Berkeley.[24] It is possible to illustrate the differences in funding level by the size of the dot used to designate the institution, as in figure 1-2.

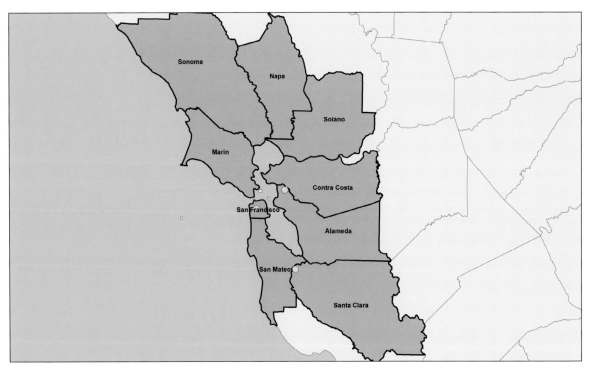

Figure 1-1. The San Francisco Bay Area with the locations of three research universities noted in yellow circles: the University of California, San Francisco (left); the University of California, Berkeley (top); and Stanford University (bottom). Courtesy of US Census.

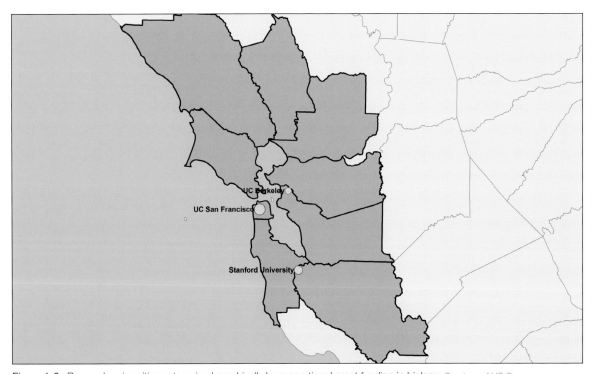

Figure 1-2. Research universities categorized graphically by proportional grant funding in biology. Courtesy of US Census.

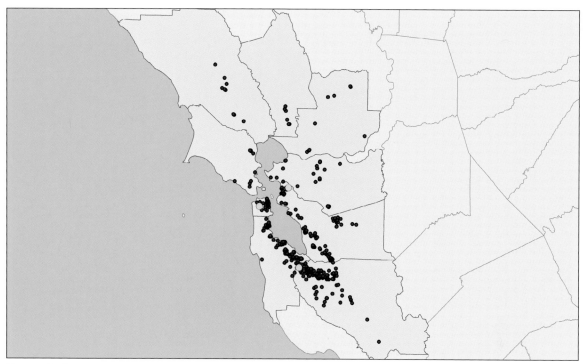

Figure 1-3. Geographical distribution of biotechnology firms (brown dots) in relation to the three research universities in San Francisco Bay Area. Courtesy of US Census and BayBio.

Since GIS ties map graphics to tabular attributes, it is a straightforward matter to assign a different-sized symbol to each university using the numerical value for funding linked to each campus. ArcGIS can do this proportional sizing automatically.

The data on biotechnology firms include street addresses. An information layer can be added to this map showing the locations of the 564 biotech firms in the nine-county area. This is illustrated in figure 1-3.

The locations of firms (based on street addresses) are converted to points on the map using geocoding—associating map locations with addresses—creating a digital "pin map." Geocoding can also be based on less precise locational information, such as ZIP Codes, or precise information, such as latitude/longitude coordinates. (Latitude/longitude coordinates might be used in cases where no street address is relevant, for example, in an area newly designated for development which does not yet have street addresses.) Geocoding is discussed in detail in chapter 7.

Figure 1-3 provides a visual sense of where firms are located. The spatial data that are represented visually are also accessible as data in a table since each point shown on the map has a corresponding record in the attribute table. Because all of the points on the map (as well as the underlying counties) possess spatial accuracy in GIS, it can be determined how many firms fall within the boundaries of any county, and generally, how many points fall within the boundaries of any polygon area for which there is map data (this could include a ZIP Code, a census tract, or a census block group). (See the sidebar, "Relationships among different geographies," in chapter 5.)

The goal is to identify some of the forces that account for the particular distribution of biotech firms displayed in the map. The heaviest clustering (Santa Clara, San Mateo, Alameda, and San Francisco) is in proximity to one of the three major research universities. The literature on technology transfer—the transfer of technological knowledge from universities and laboratories to businesses and consumers—suggests that proximity to a research university is one of the main attractors in high technology industry clustering. To gauge the relative strength of this

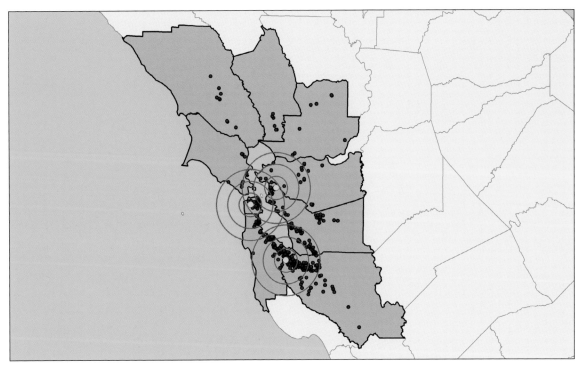

Figure 1-4. Multi-ring buffers delineating distances of five, ten, and fifteen miles from research universities. Courtesy of US Census and BayBio.

attractor, data about funding levels in biotech and related areas will be used (with data from the National Science Foundation, the National Institutes of Health, and the California Institute for Regenerative Medicine).

The map shows that the majority of biotech firms are near one of the research universities. A tool in ArcGIS can be used to measure this more exactly. *Buffers* can be created (at distances determined by the user) around each research campus and can determine the number of biotech firms contained within each buffer. ArcGIS provides a tool to create multi-ring buffers, which are concentric rings defined for specified distances from a particular location. The multi-ring buffers will be created at distances of five, ten, and fifteen miles from each of the campuses, as shown in figure 1-4. The Buffer tool used to do this in ArcGIS is one of hundreds of such *geoprocessing* tools.

The buffers are "as the crow flies" distances, not based on actual roadways and travel times. Esri provides an extension for its software called ArcGIS Network Analyst, which can help produce distances based on real-world travel paths. This is discussed in chapters 5 and 7. Drive-time areas can also be determined using Business Analyst. Chapter 2 discusses an economic development example using drive-time areas.

Since the fundamental power of GIS is to link maps with databases, the number of firms within each buffer area for each university can be computed after looking at the data that lie behind the map in figure 1-4. To construct this table, an ArcGIS tool called Intersect is used to create data in a GIS program that would be difficult or impossible to create otherwise. These unique data creation capabilities of GIS are discussed more fully in chapter 8.

Table 1-1. Number of biotech firms at each distance — five, ten, and fifteen miles — from the three research universities in the San Francisco Bay Area (counts are based on applying multi-ring buffers around research universities)

Distance from	Stanford	Berkeley	UCSF	Percent of firms
5 miles	102	34	32	29.79
10 miles	203	60	97	57.80
15 miles	318	98	146	73.58

Chapter 8 will explore spatial patterns more fully. ArcGIS provides many tools, including spatial statistical tools, for analyzing spatial patterns. One of these tools is called the Standard Deviational Ellipse, which will be discussed in chapter 8. At this point, it suffices to understand that this tool determines a directional orientation to a distribution of points—like the locations of our biotechnology firms. A casual observation of figure 1-3 shows firms tending to hug both sides of San Francisco Bay. Can this observation be confirmed statistically?

The Standard Deviational Ellipse tool creates an elongated circle around a group of points. If the figure is a circle, then the group of points implies no direction. The tool is called a Standard Deviational Ellipse because its size and location are based on the statistical concept of standard deviation (as discussed in chapter 8). An ellipse of one standard deviation is shown in figure 1-5 for each of the nine counties of the San Francisco Bay Area.

What can account for this unusual distribution? Adding a layer to represent roads reveals the answer. This is shown in figure 1-6.

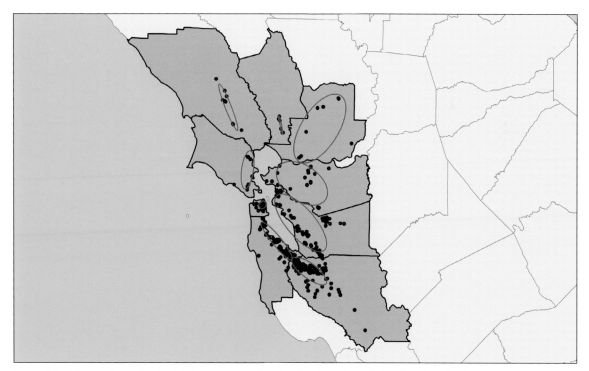

Figure 1-5. Ellipses of one standard deviation around biotechnology firms in each San Francisco Bay Area county. Courtesy of US Census and BayBio.

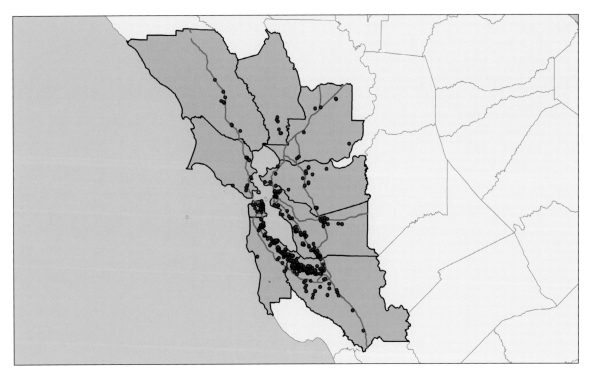

Figure 1-6. Biotechnology firms and major highways in the San Francisco Bay Area. Courtesy of US Census and BayBio.

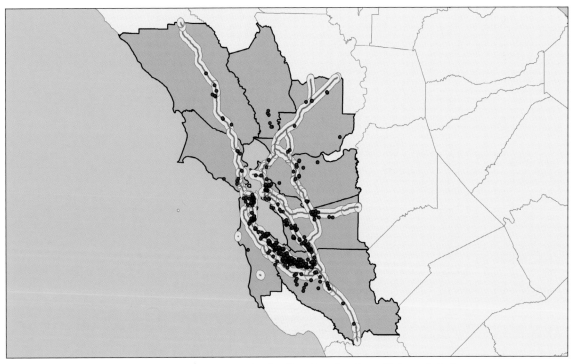

Figure 1-7. Biotechnology firms and major highways in the San Francisco Bay Area with two-mile buffers. Courtesy of US Census;
BayBio.

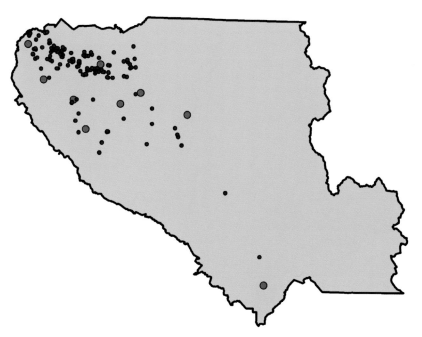

Figure 1-8. Institutions (red dots) granting associates and bachelor's degrees, and biotech firms (blue dots) in the Santa Clara County. Courtesy of US Census and BayBio. Data and assistance provided by the California Community College GIS Collaborative and the California Community College System Office.

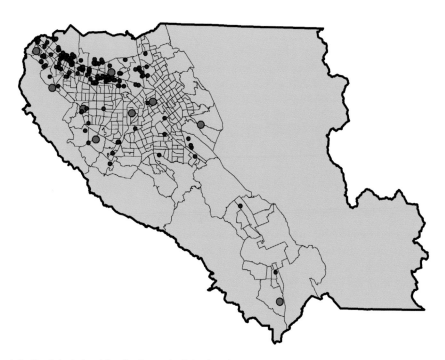

Figure 1-9. Spatial relationship of colleges (red dots) and biotechnology firms (blue dots) to Santa Clara County census tracts. Courtesy of US Census and BayBio. Data and assistance provided by the California Community College GIS Collaborative and the California Community College System Office.

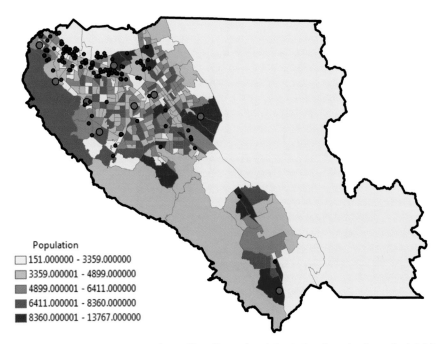

Population
- ☐ 151.000000 - 3359.000000
- ▨ 3359.000001 - 4899.000000
- ▦ 4899.000001 - 6411.000000
- ▩ 6411.000001 - 8360.000000
- ■ 8360.000001 - 13767.000000

Figure 1-10. Population distribution in Santa Clara County in relation to location of colleges (red dots) and biotechnology firms (blue dots). Courtesy of US Census and BayBio. Data and assistance provided by the California Community College GIS Collaborative and the California Community College System Office.

It is now obvious that the transportation infrastructure has a great deal to do with the location of biotechnology firms. Figure 1-6 also is notable because it contains all the types of vector data: points (biotechnology firms), lines (the highways), and polygons (the counties). The Buffer tool can be used for the lines representing highways, just as it did for the points representing research universities.

Figure 1-7 shows the two-mile buffers around the highways. The same calculations made earlier for firms within some distance of a research university can also be made for firms within some distance of a highway.

If proximity to a research university was the only consideration in firm location, it would make sense for all biotech firms to be located in the immediate vicinity of such institutions. However, biotech firms employ lower-paid lab technicians who are typically recruited locally. Therefore, relative housing prices and the presence of appropriate degree programs at the level of masters, bachelors, and associates degree might be significant factors that influence the location of biotech firms.

The chapter now focuses on one of the Bay Area counties, Santa Clara County, which has the greatest concentration of biotech firms. Figure 1-8 shows the location of the firms (blue dots) as well as two-year and four-year colleges (red dots) in the county.

The US Census Bureau provides demographic data for relatively small geographic units called census tracts. These data will be used to connect the locations of institutions of higher education with the potential pool of students. The 341 census tracts in Santa Clara County are visible in figure 1-9.

Data available from American FactFinder (see the chapter 5 sidebar, "Know the data," on American FactFinder) are used to estimate the number of people within each census tract.[25] These quantities are represented using a *color ramp* (shading) of census tracts to produce a graduated color or **choropleth map** (figure 1-10). The darker census tracts are those with a greater absolute number of people.

Determining the effectiveness of workforce development policy in biotechnology

ArcGIS contains the tools needed to quickly and easily create choropleth maps and numerous other display techniques such as **graduated symbols** (used to create the varying sizes of dots in figure 1-2), dot density maps (another technique for visualizing relative feature densities), and the addition of map bar graphs. ArcGIS tools can help users identify potentially underserved census tracts, that is, those with a large number of people in the target population who live too far from a college to make commuting to classes practical. This information is vital to educational planners and economic development officials in identifying programs for expansion. Firms seeking to evaluate potential sites also use information about available labor supply and the capacity of higher educational institutions to generate workers with the specific skills they require.

Summary

GIS is an essential tool in economic development analysis because of its data management and data manipulation capabilities, analytic tools, potential for reporting and collaboration, and scalability. GIS is the spatial equivalent of a spreadsheet. Many of the most routine problems that arise in economic development analysis can be done better and more quickly with GIS than without it—and some problems can be done only with GIS.

GIS is a natural tool for analyzing the spatial implications of economic development policies. It enables analysts to manipulate a large amount of detailed data, generate data that would not otherwise be available (for use in a GIS or statistical application), and frame problems in the context of the sophisticated economic theories that underpin economic development analysis.

This chapter focused on the *results* of using just a handful of ArcGIS tools to support an economic development analysis of industry clustering and workforce development. This chapter explained and used the economic concepts of externalities, scale economies, mobility, and related ideas to interpret the results of our GIS analysis. This interpretation was based on informal measures. More formal measures involving indexes and statistics are introduced in chapters 8 and 9.

Though some people think that GIS is difficult to use, only seven of its tools were used to complete the analysis in this chapter. As is typical in an economic development analysis, many sources of data were needed. That was the most difficult part of doing the analysis. The next chapter discusses how users can dispense with most of this burdensome data gathering and preparation.

Appendix: Data sources for the GIS application in this chapter

Census data
- American FactFinder data on population at the census tract level
- Census tracts
- Census block groups
- US states

Research funding levels
- National Science Foundation (NSF)
- National Institutes of Health (NIH)
  - National Center for Biotechnology Information
- Centers for Disease Control (CDC)

- California Institute for Regenerative Medicine (CIRM), also known as the California Stem Cell Institute
- Food and Drug Administration

Addresses: Research institutions (Carnegie classification: Research I Institutions)
Addresses: Teaching institutions (Carnegie classification: Research II Institutions)
Addresses: Community colleges
Data on enrollment and degree programs:
- California Postsecondary Education Commission
- National Center for Education Statistics
Addresses and characteristics of biotechnology firms: BayBio

Notes

[1] The Silicon Valley is an inexact geographic designation encompassing the City of San Jose, surrounding Santa Clara County, and areas to the north and south.

[2] Real gross domestic product is an aggregate measure of economic activity generally computed at the state or national level. Gross domestic product and related measures of aggregate economic activity are computed by the Department of Commerce, Bureau of Economic Analysis. Real GDP eliminates the effects of inflation, whereas nominal or current GDP does not.

[3] R. E. Sieber, "Spatial Data Access by the Grassroots." *Cartography and Geographic Information Science* 34(1) (2007): 47–62.

[4] R. E. Sieber, "Public Participation Geographic Information Systems: A Literature Review and Framework." *Annals of the American Association of Geographers* 96(3) (2007): 491–507.

[5] A wide variety of taxonomies of causes of economic development. See Emil E. Malizia and Edward J. Feser, *Understanding Local Economic Development* (New Brunswick, NJ: Center for Urban Policy Research: Rutgers, 1999); provides a good survey of the variety of economic development theories.

[6] J. M. Keynes, *The General Theory of Employment, Interest, and Money* (London: Macmillan and Co. Ltd., 1936).

[7] George A. Akerlof and Robert J. Shiller, *Animal spirits: How Human Psychology Drives the Economy, and Why it Matters for Global Capitalism* (Princeton, NJ: Princeton University Press, 2010).

[8] There are numerous applications of social network analysis. See Thomas W. Valente, *Social Networks and Health* (Oxford: Oxford University Press, 2010).

[9] Simon Thompson, "Data Sources and Assessment Tools for Entrepreneurial Development." Presented to IEDC, *Entrepreneurial and Small Business Development Strategies* (February 2010).

[10] Masahisa Fujita and Jacques-Francois Thisse, *Economics of Agglomeration: Cities, Industrial Location, and Regional Growth* (Cambridge: Cambridge University Press, 2002).

[11] See Andres Almazan, et al., "Financial Structure, Liquidity, and Firm Locations," NBER Working Paper 13660 (Cambridge, MA: National Bureau of Economic Research, 2007).

[12] See Adam B. Jaffe, Manuel Trajtenberg, Rebecca Henderson, "Geographic Localization of Knowledge Spillovers as Evidenced by Patent Citations," *Quarterly Journal of Economics* 108(3) (1993): 577–98; C. Antonelli, "Collective Knowledge Communication and Innovation," *Regional Studies* 34(6) (2000): 535–47; Piero Morosini, "Industrial Clusters, Knowledge Integration, and Performance," *World Development* 32(2) (2004): 305–326; Christopher H. Wheeler, "Do Localization Economies Arise From Human Capital Externalities?" *Annals of Regional Science* 41(1) (2007): 31–50; and Richard Florida, *Who's Your City?* (New York: Basic Books, 2008).

[13] Paul Benneworth and Gert-Jan Hospers, "The New Economic Geography of Old Industrial Regions," *Environment and Planning C* 25(6) (2007): 779–802.

[14] Jaffe, Trajtenberg, and Henderson, "Geographic Localization of Knowledge Spillovers as Evidenced by Patent Citations," 577–98.

[15] We use the six-digit NAICS (North American Industry Classification System) Code.

[16] The year 2006 is the most recent year for which data are available by six-digit NAICS Code.

[17] See appendix.

[18] Benneworth and Hospers, "The New Economic Geography of Old Industrial Regions," 779–802; Roland Andersson, Quigley, and Mats Wilhelmson, "University Decentralization as Regional Policy: The Swedish Experiment," *Journal of Economic Geography* 4 (2004): 371–388.

[19] "Remarks by the President on the American Graduation Initiative, Macomb Community College. Warren, Michigan," by Barack Obama, July 14, 2009. See http://www.whitehouse.gov/the_press_office/Remarks-by-the-President-on-the-American-Graduation-Initiative-in-Warren-MI/.

[20] Council of Economic Advisors, "Preparing the Workers of Today for the Jobs of Tomorrow." (July 2009).

[21] Council of Economic Advisors, "Preparing the Workers of Today for the Jobs of Tomorrow." (July 2009), p. ii.

[22] See "Recovery.gov Tracking the Money," http://www.recovery.gov/?q=content/act.

[23] Claudia Goldin and Lawrence F. Katz, *The Race between Education and Technology.* (Cambridge: Harvard University Press: 2010).

[24] The funding level for medicine and biology is the sum of funding from the National Institutes of Health (NIH), the Biology Program of the National Science Foundation, and funding from the California Institute for Regenerative Medicine (CIRM—the California "Stem Cell Institute") for the most recent year for which data were available.

[25] The US Census is available online at http://factfinder2.census.gov/faces/nav/jsf/pages/index.xhtml.

2

Economic development analysis using Esri Business Analyst

Objectives

- Describe the application of Esri Business Analyst Online to economic development
- Describe the application of Esri Business Analyst Desktop to economic development

On a Monday morning, a supervisor asks for a meeting and says she needs an analysis of two possible sites for a biotech firm. She wants a comparison based on the educational attainment of the population, housing prices, and incomes at various distances around in each location. She also wants a report that can be sent out by the close of business *today*.

This chapter explains how Esri Business Analyst will be used to meet this deadline. Business Analyst will also be used to address some of the same issues of industry clustering and workforce development in biotechnology that were raised in the first chapter. Additionally, Business Analyst will be applied to other geographic areas and industries to compare drive-time areas, rings, and ZIP Codes. A related issue—comparing census data available for different geographies—is discussed in chapter 5 in the sidebar, "Relationships among different geographies."

Business Analyst refers to two similar applications: **Business Analyst Online (BAO)** and **Business Analyst Desktop (BA Desktop).** As a web-based application, BAO does not require software to be installed on a computer, whereas BA Desktop as a specialized extension works in conjunction with ArcMap and is launched within ArcMap.

The most onerous tasks in economic development—acquiring and manipulating data—are greatly simplified in BAO and BA Desktop. Both applications feature access to very large datasets of public and proprietary data and tools for manipulating data. Both applications also offer immediate access to updated demographic and industry data, forecasts of demographic and industry data,[1] immediate geocoding, and a host of tools that simplify common tasks.

For example, BAO and BA Desktop can easily determine drive-time areas—areas that can be reached in a given amount of time—as well as straight-line or "as the crow files" areas (rings). Both applications also provide access to Esri's Tapestry Segmentation data (see the sidebar, "Tapestry Segmentation data and methodology," later in this chapter). Both also produce high-quality reports, and the data and maps generated from either program can be exported for use in other software applications. BAO provides a simple workflow template; BA Desktop provides several templates for a variety of typical economic development tasks. Chapter 4 illustrates the use of the BA Desktop workflow in selecting a site for a retail store.

The main trade-off between the programs is greater ease of use (for BAO) versus greater flexibility in employing larger and more varied user-supplied supplementary datasets (for BA Desktop). The following table gives a thumbnail comparison of the two applications. BAO is limited in the number of addresses that can be geocoded at one time and in the amount and type of user-supplied supplementary data that can be employed in an analysis.

Both BAO and BA Desktop have a functionality called **SmartMap Search,** which lets users search the entire Business Analyst database for sites satisfying multiple characteristics. The search capabilities are illustrated later in this chapter.

Table 2-1. Comparison of Esri Business Analyst Online and Esri Business Analyst Desktop

Features	Business Analyst Desktop	Business Analyst Online
No software installation		Y
Ease of use		Y
Geocoding	immediately implementable; unlimited addresses	immediately implementable; up to 100 addresses
Driving distance analysis	Y	Y
Site selection tools	Y	Y
Analytic tools	rank markets, geographic trade areas, find similar successful locations, find competitors, proximity to competitors, measure cannibalization, Huff Model of market potential, evaluate store performance using trade areas, find areas and target new customers	
Updated demographic data	Y	Y
Updated industry data	Y—SIC, NAICS Codes	Y—SIC, NAICS Codes
Forecast data	Y	Y
Tapestry Segmentation data	Y	Y
Community Coder	Y	
Geographies	census tracts, counties, MSAs, CBSAs, ZIP Codes	census tracts, counties, MSAs, CBSAs, ZIP Codes
Report writing	Crystal Reports	PDF, Excel
User-supplied supplementary data (e.g., customer data)	unlimited	100 records (.dbf or Excel); 100 contiguous polygons
Workflow templates	Y	Y

Business Analyst Online

Referring to the example at the start of the chapter, the supervisor who asked for the site analysis has selected San Jose, California, and Austin, Texas, as two possible locations for a biotechnology firm. These cities highlight the comparison capabilities of BAO. San Jose and Austin are medium-sized cities near major research universities. (The main campus of the University of Texas is located in Austin. Chapter 1 noted the proximity of San Jose to three research universities in the San Francisco Bay Area: Stanford, Berkeley, and the University of California, San Francisco.)

Because of the short timeline to make a final report, BAO is a good choice to do the analysis. With BAO, users can access the required data, make comparisons, and generate final reports—all with a few mouse clicks. All of the maps, tables, and graphs for both examples below were generated in thirty to forty-five minutes using BAO.

Launching Business Analyst Online

To make the analysis and have it ready for the supervisor, users can access BAO by going to the Esri BAO website and entering their usernames and passwords (figure 2-1). For these examples and others in the book, the interface you see on your screen may vary slightly depending on the version of software used.

Business Analyst Online workflow

The BAO welcome page has three main tabs in addition to the Home tab. These tabs—Select Location, Get Reports, and Research Market—offer a variety of alternatives. With BAO, users also can pinpoint a specific location, upload files, draw on a map, and select a variety of geographies (census tracts, county subdivisions, Congressional districts, and so forth). Users can also search for locations with specific characteristics.

Figure 2-1. Login page for Business Analyst Online.

Example 1: Comparing areas for biotechnology potential: Santa Clara County, California, and Travis County, Texas

By using the Select Locations tab, users can locate both Santa Clara County (which contains the city of San Jose) and Travis County (which contains the city of Austin). BAO is used to render maps of each of these counties, as illustrated in figure 2-2.

BAO can help users determine some basic facts about each place and also obtain detailed information about the types of industry and the characteristics of their respective workforces. It is important for the client to know how these areas differ in terms of education and skills required for the biotechnology firm. So, the first step is to compare the two areas on population, housing, and income. These are the variables included under "standard reports" in the BAO "Get Reports" tab. The population comparison is given in figure 2-3.

BAO offers forecasts for many variables five years into the future in addition to current data. Figure 2-4a shows a comparison of 2010 income statistics, and figure 2-4b shows a comparison of income statistics forecast for the two counties in 2015.

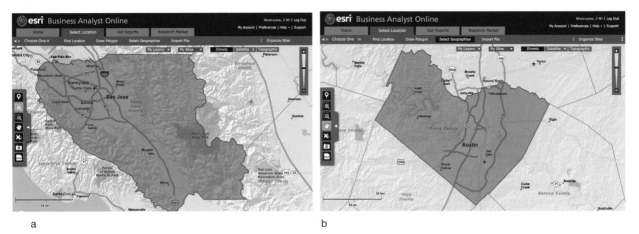

a   b

Figures 2-2a and b. Santa Clara County, California, with the main tabs of BAO highlighted (a); and Travis County, Texas (b).

Data displayed in screenshots of Esri Business Analyst are courtesy of Esri; US Census Bureau; Infogroup; Bureau of Labor Statistics; Applied Geographic Solutions, Inc.; Directory of Major Malls, Inc.; GfK MRI; and Market Planning Solutions, Inc.

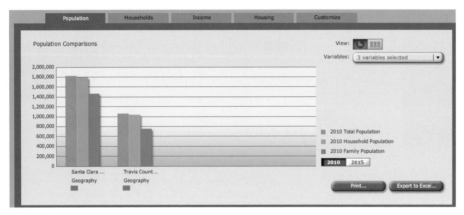

Figure 2-3. Population comparison between Santa Clara and Travis Counties. Data displayed in screenshots of Esri Business Analyst are courtesy of Esri; US Census Bureau; Infogroup; Bureau of Labor Statistics; Applied Geographic Solutions, Inc.; Directory of Major Malls, Inc.; GfK MRI; and Market Planning Solutions, Inc.

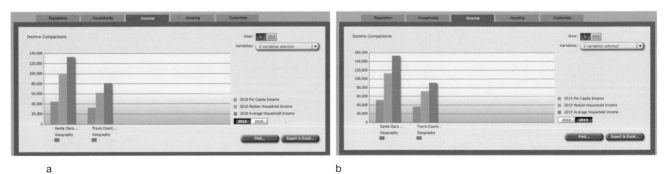

a   b

Figure 2-4a and b. Comparison of 2010 incomes of Santa Clara and Travis Counties (a) and comparison of incomes of Santa Clara and Travis Counties, forecast for 2015 (b). Data displayed in screenshots of Esri Business Analyst are courtesy of Esri; US Census Bureau; Infogroup; Bureau of Labor Statistics; Applied Geographic Solutions, Inc.; Directory of Major Malls, Inc.; GfK MRI; and Market Planning Solutions, Inc.

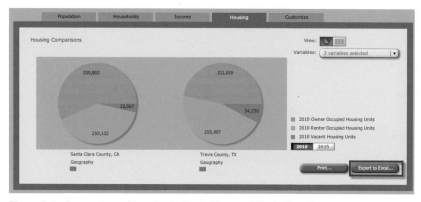

Figure 2-5. Comparison of housing in Santa Clara and Travis Counties. Data displayed in screenshots of Esri Business Analyst are courtesy of Esri; US Census Bureau; Infogroup; Bureau of Labor Statistics; Applied Geographic Solutions, Inc.; Directory of Major Malls, Inc.; GfK MRI; and Market Planning Solutions, Inc.

The basic housing comparison is given in figure 2-5.

The data underlying these graphs (and any other graphs and tables generated by BAO) can be downloaded as Excel files; the "Export to Excel" button is highlighted in the lower right corner of figure 2-5.

BAO can also be used to create custom reports that are responsive to specific client needs. BAO provides a wealth of immediately accessible data. To address the site analysis requirements in this example, several educational attainment and workforce variables can be selected, as illustrated in figure 2-6.

Figure 2-6. Selecting variables for a custom comparison report in BAO.

These variables were selected from one of six categories, in this case, "Demographics." The other categories are: Business, Consumer Spending, Market Potential, Retail Marketplace, and Tapestry Segmentation. There is also an option to select variables from all categories.

Figure 2-7 gives the results of the selection. The graph indicates the number of those sixteen years and older employed in life sciences, and the numbers of those twenty-five years or older whose highest degree is an associate, bachelors, or graduate, respectively. The age categories for occupational and educational attainment data are determined by the US Census Bureau.

Figure 2-7 is presented in absolute numbers of people in each category, but BAO also lets users present the same data as percentages of the total.

Because the task is to determine the best place for a biotechnology firm in one of these counties, it is important to compare the existing industry and employment structure in each area. The Custom Reports tab allows the

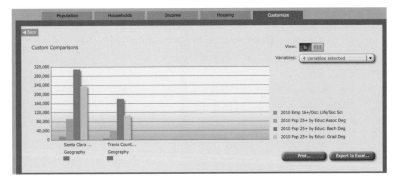

Figure 2-7. Educational attainment of population twenty-five years and older and occupational classification of population sixteen years and older: comparison of Santa Clara and Travis Counties. Data displayed in screenshots of Esri Business Analyst are courtesy of Esri; US Census Bureau; Infogroup; Bureau of Labor Statistics; Applied Geographic Solutions, Inc.; Directory of Major Malls, Inc.; GfK MRI; and Market Planning Solutions, Inc.

selection of North American Industry Classification System (NAICS) Codes (see the chapter 1 sidebar) that focus on industries of interest; both the number of businesses in a particular NAICS Code and employment by NAICS Code can be selected. The selection of variables is illustrated in figure 2-8.

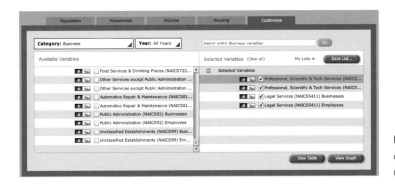

Figure 2-8. Selection of NAICS Codes for comparison of industry structure in Santa Clara and Travis Counties.

Figure 2-9 shows the result of the NAICS comparison.

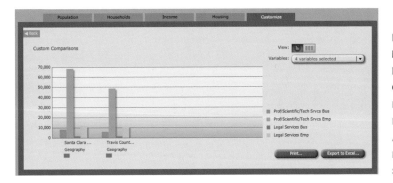

Figure 2-9. Comparison of number of businesses and employment by selected NAICS Code for Santa Clara and Travis Counties. Data displayed in screenshots of Esri Business Analyst are courtesy of Esri; US Census Bureau; Infogroup; Bureau of Labor Statistics; Applied Geographic Solutions, Inc.; Directory of Major Malls, Inc.; GfK MRI; and Market Planning Solutions, Inc.

Success! These comparisons offer enough information to make an informed site selection.

Example 2: Comparing three areas for solar panel production potential

But success does have its price. In this scenario, the work on the biotech siting project has led to more work, this time a comparison of three small cities as potential locations for a solar panel manufacturing facility. The report needs to be done immediately to compare the three selected candidates—Elkhart, Indiana; Kalamazoo, Michigan; and Fort Myers, Florida. The sites displayed in each of the cities in the figures below were chosen at random.

BAO will be used to compare the three small cities as possible sites for a new solar panel production facility. Small cities often attract business because they tend to have lower labor costs than larger urban areas. They are often the focus of federal and state efforts to spur economic development, and financial incentives may be offered to firms that locate in especially distressed areas, for example, in areas where the unemployment rate is 50 percent or more above the national average.

An experienced labor force is a key consideration in choosing among the cities. Because executives and managers will have to relocate to the cities, the overall character of the cities is also an important consideration.

Relying only on public data sources for economic development analysis, especially for industry, can be problematical. The main source of detailed public data about industries is the Economic Census, which is often years out of date, and the lowest geographic level at which these data are available is the ZIP Code. While ZIP Codes might provide an acceptable level of geographic specificity in a large urban area, they are often inadequate for analysis in smaller cities, which might have few (or even one) ZIP Codes, and typically spill into rural areas. Business Analyst, in contrast, provides immediate access to nonpublic data below the ZIP Code level.

The three cities mentioned above are all small and relatively isolated from urban centers. They differ on several dimensions, including the presence of universities. Elkhart is close to Indiana University in South Bend and the University of Notre Dame, both of which are in adjacent St. Joseph County. Kalamazoo is home to Western Michigan University, and Fort Myers is home to Florida Gulf Coast University.

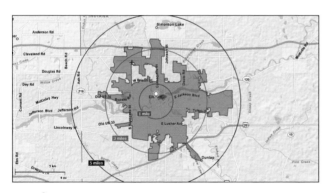

a

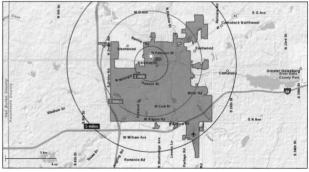

b

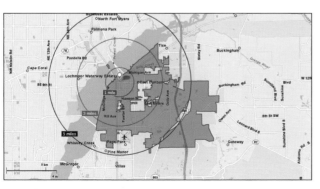

c

Figures 2-10a, b, and c. Elkhart, Indiana (top left), Kalamazoo, Michigan (right), and Fort Myers, Florida (lower left). Data displayed in screenshots of Esri Business Analyst are courtesy of Esri; US Census Bureau; Infogroup; Bureau of Labor Statistics; Applied Geographic Solutions, Inc.; Directory of Major Malls, Inc.; GfK MRI; and Market Planning Solutions, Inc.

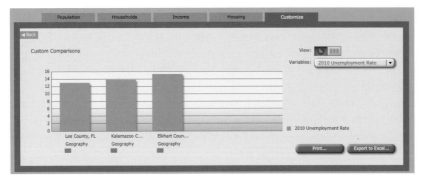

Figure 2-11. The unemployment rate (2010) in Elkhart, Kalamazoo, and Fort Myers. Data displayed in screenshots of Esri Business Analyst are courtesy of Esri, US Census Bureau, Infogroup, Bureau of Labor Statistics, Applied Geographic Solutions, Inc., Directory of Major Malls, Inc., GfK MRI, and Market Planning Solutions, Inc.

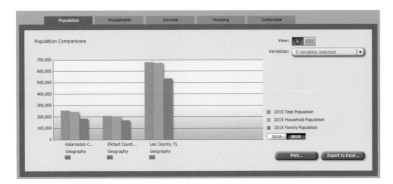

Figure 2-12a. Comparison of populations in Elkhart, Kalamazoo, and Fort Myers. Data displayed in screenshots of Business Analyst are courtesy of Esri; US Census Bureau; Infogroup; Bureau of Labor Statistics; Applied Geographic Solutions, Inc.; Directory of Major Malls, Inc.; GfK MRI; and Market Planning Solutions, Inc.

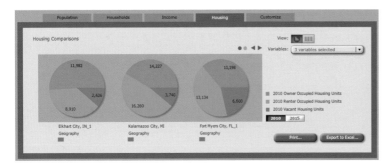

Figure 2-12b. Housing comparison. Data displayed in screenshots of Esri Business Analyst are courtesy of Esri; US Census Bureau; Infogroup; Bureau of Labor Statistics; Applied Geographic Solutions, Inc.; Directory of Major Malls, Inc.; GfK MRI; and Market Planning Solutions, Inc.

Figure 2-12c. Comparison of average incomes. Data displayed in screenshots of Business Analyst are courtesy of Esri; US Census Bureau; Infogroup; Bureau of Labor Statistics; Applied Geographic Solutions, Inc.; Directory of Major Malls, Inc.; GfK MRI; and Market Planning Solutions, Inc.

The cities are all relatively compact, as indicated by the maps in figure 2-10a, b, and c. Rings of one mile, three miles, and five miles are included to provide a reference. These maps were generated by BAO as part of the workflow template for generating analyses and reports.

Each of these cities had an unemployment rate significantly higher than the national average unemployment rate of 9.7 percent in 2010 (see figure 2-11).[2]

In BAO, the "run standard reports" option can compare the cities on the basic dimensions of population, housing, and income. The comparisons are shown in figures 2-12a, 2-12b, and 2-12c.

Because the goal is to choose a location for a manufacturing facility, the report will focus on the occupational and employment backgrounds of the population. (These specific details are easily obtained by creating a custom report in BAO but are not part of the standard report.) Custom reports can be created to select relevant variables (see figure 2-13).

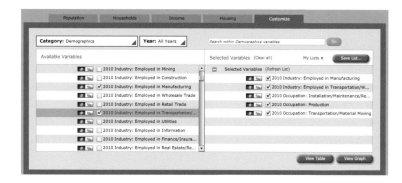

Figure 2-13. Selecting occupational and employment variables relevant to solar panel manufacturing.

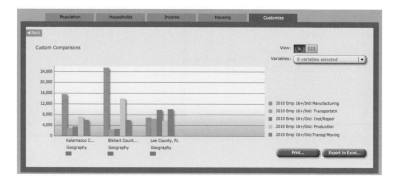

Figure 2-14. Occupational and employment backgrounds of the respective labor forces. (County-level data). Data displayed in screenshots of Esri Business Analyst are courtesy of Esri; US Census Bureau; Infogroup; Bureau of Labor Statistics; Applied Geographic Solutions, Inc.; Directory of Major Malls, Inc.; GfK MRI; and Market Planning Solutions, Inc.

The employment levels shown in figure 2-14 are in absolute numbers, but BAO can also represent the data as percentages of the local labor force. Clearly, Elkhart is the area with the highest employment in manufacturing and production. This addresses the first question concerning the workforce characteristics of the cities.

Analysis of demographic information

The "character of the community" is just as important as the characteristics of the labor force in determining the suitability of a prospective site. Managers and executives will live in the community, and different corporate cultures mesh better with some communities than others. BAO can be used to access the Tapestry Segmentation database, which is frequently used to identify the suitability for a specific retail market; its data can also be used to get a sense of the community.

Tapestry Segmentation data and methodology

The Tapestry Segmentation database is a nationwide database developed by Esri. The primary purpose of the database is to create a categorization (taxonomy) of households for purposes of identifying market segments. Tapestry Segmentation is based on a "market segmentation" model and methodology. Households in each community are identified with one of sixty-five groups based on lifestyle and degree of urbanization. The taxonomy into sixty-five groups is based on an analysis of what drives consumer purchases. Households in the same group have similar demands for goods and services. The primary use of the Tapestry Segmentation dataset is to identify markets for categories of goods that would be purchased by different households. For example, households consisting of retired persons living in rural areas demand different goods and services than those of single people living in high-density urban areas.

The Tapestry Segmentation data can be accessed for geographies, such as cities or ZIP Codes. The data will identify the predominant groups in the selected geography.

Although the primary purpose of the Tapestry Segmentation dataset is to facilitate marketing, the data can also be used to infer something about the "character of the community." The predominant Tapestry Segmentation groups will be identified for geographies selected by the user.

For example, the segment "Laptops and Lattes" is described as follows: "With no home ownership or child-rearing responsibilities, residents of *Laptops and Lattes* neighborhoods enjoy single life in the big city. Most households are singles who live alone or with a roommate. The average household size remains constant at 1.8. Although this segment is slowly increasing, it is maturing and

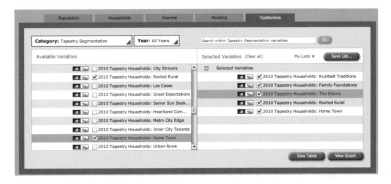

Figure 2S-1. Selected Tapestry Segmentation variables.

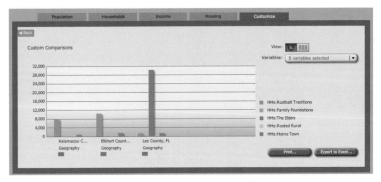

Figure 2S-2. Comparing cities using selected Tapestry Segmentation variables. Data displayed in screenshots of Esri Business Analyst are courtesy of Esri; US Census Bureau; Infogroup; Bureau of Labor Statistics; Applied Geographic Solutions, Inc.; Directory of Major Malls, Inc.; GfK MRI; and Market Planning Solutions, Inc.

continued ➡

diversifying more quickly. The median age is 38.7 years. Although most of the population is white, Asians represent 10.4 percent of the total population."[3]

In figure 2S-1 some tapestry variables that might help define the character of the three cities have been selected.

Figure 2S-2 shows the results of comparing the three small cities based on the selected Tapestry Segmentation variables (at the county level).

Selecting the relevant variables from the large number available in the Tapestry Segmentation dataset depends on the focus of the analysis.

Drive-time areas, rings, and ZIP Codes

Determining drive-time areas is very important in economic development analysis. A retail outlet, for example, cannot be too far from its customers, and it needs to be relatively close to a workforce with the appropriate skills. If the industry is one that benefits from *economies of localization,* based for example on the need for face-to-face contact with colleagues in other firms (see the discussion in chapter 1), it needs to be in relative proximity to other firms in the same industry.

Generally, "too far" should be construed in terms of how long it will take to drive to a location, rather than the straight-line distance. Neither employees nor customers are crows! BAO can easily generate drive-time areas. Figures 2-10a to 2-10c displayed one-mile, three-mile, and five-mile rings for each of the small cities. Figure 2-15 displays drive-time areas for the city of Elkhart.

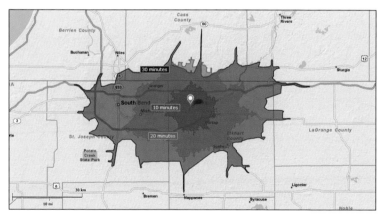

Figure 2-15. Drive-time areas (ten minutes, twenty minutes, and thirty minutes) for Elkhart, Indiana. The irregular shape of the drive-time areas is dictated by the locations and capacities of roads and highways. The "spikes" generally correspond to highways that allow for faster travel speeds and therefore greater distances between work and home. Data displayed in screenshots of Esri Business Analyst are courtesy of Esri; US Census Bureau; Infogroup; Bureau of Labor Statistics; Applied Geographic Solutions, Inc.; Directory of Major Malls, Inc.; GfK MRI; and Market Planning Solutions, Inc.

Data can also be displayed by ZIP Code. Elkhart overlaps three ZIP Codes (46514, 46516, and 46517). The site identified in Elkhart lies in ZIP Code 46516. This ZIP Code is shown in figure 2-16 with an overlay of five-minute, ten-minute, and fifteen-minute drive-time areas.

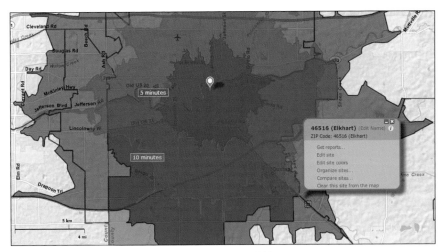

Figure 2-16. ZIP Code 46516 and drive-time areas for Elkhart. Data displayed in screenshots of Esri Business Analyst are courtesy of Esri; US Census Bureau; Infogroup; Bureau of Labor Statistics; Applied Geographic Solutions, Inc.; Directory of Major Malls, Inc.; GfK MRI; and Market Planning Solutions, Inc.

BAO makes the drive-time areas and the ZIP Codes immediately accessible to the user. This is important for economic development and business users in part because business mailings often target ZIP Codes. Overlaying the drive-time areas can determine which ZIP Codes should be the focus of mailings.

Figure 2-16 illustrates drive-time areas in Elkhart. The ZIP Code containing the potential site (46516) is overlaid on the drive times to illustrate that analyses based on ZIP Code-level data may poorly reflect drive times. This is especially true in areas like small cities where population and density is lower and ZIP Codes cover a larger area. ZIP Codes are really administrative areas for the post office. They generally cover small geographic areas in densely populated places (Manhattan, San Francisco) and larger geographic areas in less densely populated places (small towns like Elkhart, Kalamazoo, and Fort Myers).

As noted earlier, the ZIP Code level is the smallest geography for which the Economic Census data on industries are available. Figure 2-16 shows that the ZIP Code does not bear any particular relationship to the drive-time areas. Accessing ZIP Code-level data alone would not offer a good approximation to proximity for purposes of commuting or for face-to-face contact with colleagues or convenience for customers. This is especially true in small cities with lower population and lower density where ZIP Codes tend to cover larger areas.

Geocoding in Business Analyst Online

Geocoding (associated locations on a map with addresses or longitude and latitude) can be done using BAO in any of three ways:
- Inputting a street address
- Inputting a latitude and longitude
- Clicking on a point on the map displayed in BAO (this is called "reverse geocoding")

Either of the first two methods can be done on a case-by-case basis (entered individually) or by uploading up to one-hundred street addresses or latitude/longitude combinations as a spreadsheet or .dbf file format. Street address data are the most likely to come from directories, while latitude-longitude data arise from a GPS tracking unit.

In this case, address information has been uploaded for three of the public community colleges in Santa Clara County into BAO, which determined the five-, ten-, and fifteen-minute drive-time areas around each college. The results are displayed in figure 2-17.

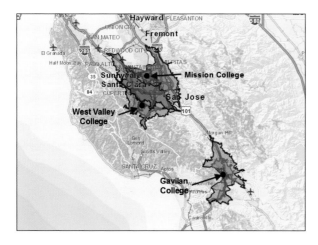

Figure 2-17. Drive-time areas (five minutes in red, ten minutes in green, and fifteen minutes in purple) around three Santa Clara County public community colleges. Data displayed in screenshots of Esri Business Analyst are courtesy of Esri; US Census Bureau; Infogroup; Bureau of Labor Statistics; Applied Geographic Solutions, Inc.; Directory of Major Malls, Inc.; GfK MRI; and Market Planning Solutions, Inc.; and data and assistance provided by the California Community College GIS Collaborative and the California Community College System Office.

BAO was then used to determine the population within each drive-time area of each college. The results, which can also be downloaded as an Excel file, are displayed in figures 2-18a and 2-18b.

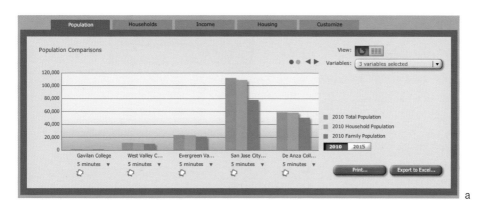

a

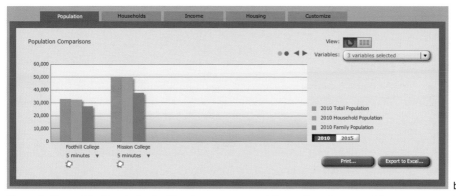

b

Figure 2-18a and b. Population in drive-time areas around Santa Clara County public community colleges (a). Population in drive-time areas around Santa Clara County public community colleges (b). Data displayed in screenshots of Esri Business Analyst are courtesy of Esri; US Census Bureau; Infogroup; Bureau of Labor Statistics; Applied Geographic Solutions, Inc.; Directory of Major Malls, Inc.; GfK MRI; and Market Planning Solutions, Inc.; and data and assistance provided by the California Community College GIS Collaborative and the California Community College System Office.

Finally, figure 2-19 shows a color-coded map for the category "Laptops and Lattes."

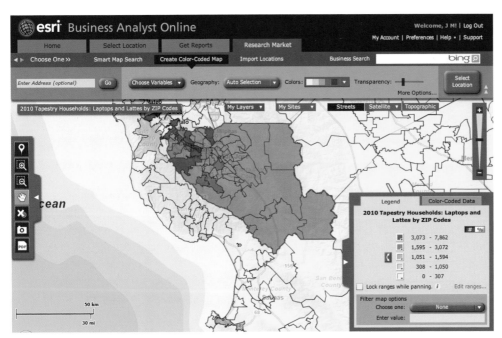

Figure 2-19. "Laptops and Lattes." Data displayed in screenshots of Esri Business Analyst are courtesy of Esri; US Census Bureau; Infogroup; Bureau of Labor Statistics; Applied Geographic Solutions, Inc.; Directory of Major Malls, Inc.; GfK MRI; and Market Planning Solutions, Inc.

The report for the three cities is now ready.

Business Analyst Desktop

BA Desktop also works seamlessly with ArcCatalog and ArcToolbox.

Why use Business Analyst Desktop?

BA Desktop is used mainly to do more extensive analysis and access a greater number of analytic tools. ArcGIS for Desktop comes with a large number of tools, including specialized statistical tools that can be used in economic development analysis (see chapter 8). The BA Desktop application comes with still more tools and a large body of data ready for use. Businesses can be selected from a database with more than 11 million listings. Businesses can be chosen by name, location, SIC (Standard Industry Classification), or NAICS Code and can be filtered as to size (employment or revenues) and structure. Furthermore, larger and more diverse user-supplied data can be incorporated in the analysis with BA Desktop. BAO and BA Desktop both provide immediate drive-time areas and geocoding.

Tools for geographic analysis and problem-solving

Esri's suite of desktop GIS software, ArcGIS, contains three primary, interlinked applications: ArcCatalog, ArcMap, and ArcToolbox. Chapter 1 explained that GIS is like a "spreadsheet for maps." The primary applications within ArcGIS for Desktop can be thought of in terms of analogous spreadsheet applications. These applications closely integrate with ModelBuilder, a feature that allows visualization of geoprocessing (see chapter 3).

The first and most foundational of the applications is **ArcCatalog,** which is used to preview and manage the datasets that are used in ArcGIS for Desktop and represent geographic features.

This is similar to some of the housekeeping functions of spreadsheets—the properties of the workbook and worksheets—such as security settings on the spreadsheet. For spreadsheets, many users never concern themselves with these housekeeping details. In GIS, some more housekeeping needs to be done using the ArcCatalog application.

ArcMap, the second primary component of the ArcGIS suite, is where map-based project work is done. Datasets are added to ArcMap's table of contents as **map layers.** Datasets can also be "streamed" directly into ArcMap via an Internet connection to a remotely served data source. For example, ArcMap offers users the ability to connect to ArcGIS Online, an Esri-hosted service that makes it simple to select from a wide variety of commonly used datasets (terrain, jurisdictional boundaries, aerial photographs, and so on) and, with one or two clicks, have these datasets added "live" to ArcMap. These remotely streamed datasets can be used alongside locally stored datasets to create detailed maps for analytical purposes.

ArcMap is used to make map documents (the name for a complete set of map layers, and assigned an .mxd file extension). Each data layer added to a map document is **georeferenced,** meaning that, ideally, all data layers (parcels, streets, water features, and so on) will be positioned correctly in the map representation in relation to their location on the earth's surface and correctly positioned relative to geographic objects represented in other data layers. Occasionally, the user must intervene to correct data layer alignment problems caused by missing or incomplete information from the data author. Additional datasets, such as spreadsheet tables or graphic files such as agency logos, can be added directly to ArcMap.

Once added into ArcMap, the layers of digital geographic data can be grouped and symbolized to form a **base map** with thematic (theme, or topic-based) datasets overlaid on top of the base map. An example of a base map would be one containing the jurisdictional boundaries of a community, roadways, water features, and appropriate labels—these features are commonly used to provide the foundation or base to many project maps. An example of a thematic dataset, added for analytical purposes to a base map, might be demographic information from the Census Bureau, such as census tracts color-coded according to a quantitative variable (also known as a choropleth map), as seen in chapter 1 with the Santa Clara County census tract data. BAO does a lot of this automatically (for example, predesigned base maps). ArcMap is used to create tailored, customized maps for analysis or maps that require large amounts of user-supplied data.

ArcMap is also used to represent the results of analyses of spatial relationships generated by statistical analysis and geoprocessing. Another core function of ArcMap is to prepare finished work for final output destined for a printed page, website, or presentation display. ArcMap generates what most people expect from a GIS program—maps—just as spreadsheets generate what most people recognize as the output of a spreadsheet—the display of words and numbers in cells arranged in rows and columns.

ArcToolbox, the third primary application in the ArcGIS suite, integrates tightly with ArcCatalog and ArcMap and contains hundreds of preprogrammed tools and functions designed to accomplish

continued

specific map-based tasks. Chapter 1 presented some of these tools, including the Intersect and Multi-ring Buffer tools. Other ArcToolbox tools are used for statistical analyses, the construction of procedural models using ModelBuilder, the conversion of one data type to another, and many more. All of the tools embrace the procedure known as **geoprocessing**, the creation of new datasets by applying tools and processes to existing datasets. ArcToolbox is like the built-in functions and data manipulation capabilities of spreadsheet programs (ModelBuilder is like a visual version of a spreadsheet "macro").

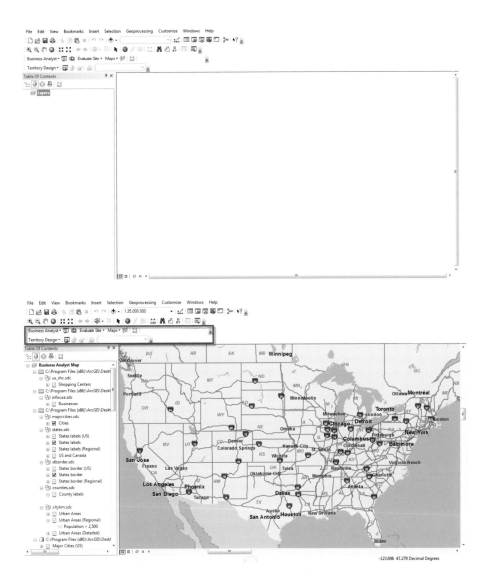

Figures 2-20 and 2-21. The default screen for ArcMap without Business Analyst (on top), and the default screen for ArcMap with BA Desktop (on bottom). Data displayed in screenshots of Esri Business Analyst in figure 2-21 are courtesy of Esri; US Census Bureau; Infogroup; Bureau of Labor Statistics; Applied Geographic Solutions, Inc.; Directory of Major Malls, Inc.; GfK MRI; and Market Planning Solutions, Inc.

The difference between ArcGIS with and without Business Analyst is apparent upon launch. Figure 2-20 shows the default (blank map) screen for ArcMap without Business Analyst, while figure 2-21 shows the default screen for BA Desktop.

Both screens have the same general structure—a menu/tool bar across the top, a left-hand table of contents pane, and a right-hand display pane (your interface may vary slightly, depending on the version of software used). The menu/tool bar lists some of the tools available for application, the table of contents pane lists the layers that are part of the map, and the display pane renders the map. In the default case for ArcMap without Business Analyst, the table of contents pane is empty, and therefore the display pane is empty—there is no content to display or render. Of course, the user can add map layers and tabular data, but the default starts from a blank map.

In contrast, upon launch the default Business Analyst has a long list of items in the table of contents pane (so many that not all are displayed because the map would be too cluttered; elements listed in the table of contents pane can be turned on or turned off by the user). Also, a set of available tools highlighted on the left-hand end of the menu/tool bar are not available in ArcMap without BA Desktop.

Consider the earlier scenario of comparing San Jose, California, and Austin, Texas, as suitable sites for biotechnology firms. Because biotechnology firms tend to cluster due to agglomeration economies, it is important to know how many firms are in each area and how they are distributed. This is readily accomplished using BA Desktop by first defining a "study area" comprising Santa Clara and Travis Counties, then determining the number of biotechnology firms within the 541711 NAICS Code. Figure 2-22 shows the result of using BA Desktop's NAICS search of the Business Analyst database for the two counties.

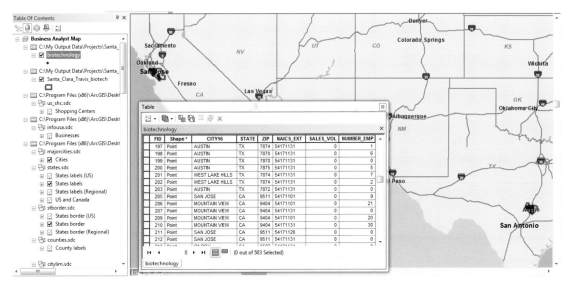

Figure 2-22. Biotechnology firms in Santa Clara and Travis Counties, represented in BA Desktop. Data displayed in screenshots of Esri Business Analyst are courtesy of Esri; US Census Bureau; Infogroup; Bureau of Labor Statistics; Applied Geographic Solutions, Inc.; Directory of Major Malls, Inc.; GfK MRI; and Market Planning Solutions, Inc.

Similarly, BA Desktop can be used to determine the number of firms associated with the potential for producing solar panels in the counties containing Kalamazoo, Elkhart, and Fort Myers. (These would be the firms with the three-digit NAICS Code 335 for electrical equipment, appliance, and component manufacturing.) Highlighted columns in figure 2-23 indicate the city in which the firm is located, the sales volume, and the number of employees.

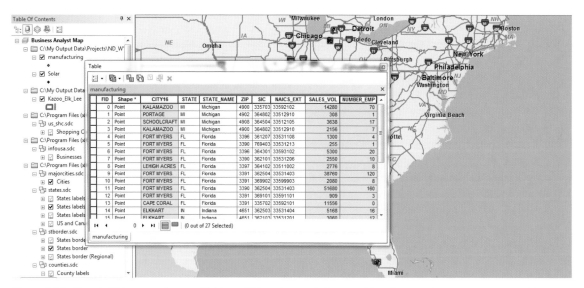

Figure 2-23. Firms in Kalamazoo County, Michigan, Elkhart County, Indiana, and Lee County, Florida, in the 335 NAICS Code.

Data displayed in screenshots of Esri Business Analyst are courtesy of Esri; US Census Bureau; Infogroup; Bureau of Labor Statistics; Applied Geographic Solutions, Inc.; Directory of Major Malls, Inc.; GfK MRI; and Market Planning Solutions, Inc.

Chapter 8 includes a more detailed discussion of the many tools available for analyzing this data, while chapter 4 explores, in depth, an application of BA Desktop to site selection. That application also illustrates more of the data and tools available in BA Desktop.

Conclusion

This chapter illustrated how BAO and BA Desktop can be used in economic development. The most onerous tasks in economic development—acquiring and manipulating data—are greatly simplified using these applications. Both feature access to a very large dataset of public and private data and tools for manipulating data.

Now that this chapter has shown how some tools of GIS *can* be used to address economic development problems, the next chapter will explain how these tools *should* be used as "best practices" in GIS for economic development analysis.

Notes

[1] The forecast data are provided by Esri's Forecasting Unit. See http://www.Esri.com/data/Esri_data/demographic.html.

[2] The national average unemployment rate through September 2010 was 9.7 percent (US Bureau of Labor Statistics).

[3] Esri, Tapestry Segmentation Reference Guide, Redlands, CA: Esri, n.d., 34.

3

GIS best practices in economic development analysis

Objective

- To introduce the Geographic Approach, which enables the following:
  - Effective GIS project management
  - Geodatabase capacity and facility
  - GIS data quality assurance
  - Automated workflows

An economic development analysis employing GIS is often a team project. Central to the success of the project is coordination and organization. This chapter looks at some tools and procedures of GIS that can promote greater organization and coordination of the team.

The job of team leader is to successfully complete a complex economic development analysis. The true measure of success may be measured in the years of service that come from the dataset and methods put in place at the very beginning of the project. This chapter focuses on what team leaders need to know about GIS tools and procedures. This would be the case for those who direct any part of a project that includes a GIS component, even if they are not conversant with GIS.

Why incorporate GIS best practices?

After finishing chapters 1 and 2, readers will want to learn more about ArcGIS software and Esri Business Analyst as powerful tools for economic development analysis and collaboration. This chapter steps back from a demonstration of these tools to consider some time-tested principles that enable efficient and organized projects based on GIS. Every field of endeavor has a set of best practices and tools that, through accumulated sharing of wisdom, enables its practitioners to pursue tasks in a manner that saves time and yields desired outcomes. This chapter presents the methods for implementing GIS projects in a straightforward and orderly manner.

In integrating GIS into a project, an important question is how to begin in a deliberate and organized way. Because it takes time to understand an advanced tool like ArcGIS, this chapter aims to provide a holistic overview of a GIS project that also gets project members up to speed quickly. This chapter also explains why meaningful economic development must incorporate spatial analysis tools and geospatial data that lie at the core of a GIS. The best practices advocated in this chapter are transferable to all GIS projects. For continuity purposes, many of these best practices will be related to the workforce development analysis conducted in chapter 1.

Adopting the Geographic Approach for GIS-based analysis

Adopting a GIS-based research methodology known as the Geographic Approach will help lay the foundation for a successful project. This problem-solving approach is a reliable method for location-based analysis and decision-making. According to Jack Dangermond, president of Esri, the Geographic Approach is

". . . a new way of thinking and problem solving that integrates geographic information into how we understand and manage our planet. This approach allows us to create geographic knowledge by measuring the earth, organizing these data, and analyzing and modeling various processes and their relationships. The Geographic Approach also allows us to apply this knowledge to the way we design, plan, and change our world."[1]

The Geographic Approach for implementing projects is likely familiar to many readers already because it calls for an organized, sequential, data-focused strategy. In the broadest sense, this strategy involves first clearly defining a location-based problem and project objectives, then obtaining the necessary datasets for analysis, performing the analysis in GIS, presenting the results, and documenting the methodology.

A deliberate GIS research strategy is essential for project success since almost all economic development analyses involve myriad variables, such as the suitability of a site for a new business (as demonstrated in chapter 2), the locations of competing businesses, or the demographic composition and spatial distribution of a target consumer market. Furthermore, ArcGIS facilitates sharing knowledge via digital maps and spatial data. It allows the project team to easily create new information and summary data for specific project needs and also create specific tools that foster the creation of scalable geographic databases. These databases grow over time as more information is acquired. Using a particular form of ArcGIS database called a file geodatabase will help users achieve these sharing and scalability goals. As the term implies, a geodatabase (*geo* and *database*) centralizes a project's geographic data.

The adoption of the Geographic Approach recognizes that spatial relationships between labor pools, markets, jurisdictions, and transportation networks are central to economic development theory and are the key to effectively solving the location-focused problems in economic development.

The five steps of the Geographic Approach

The Geographic Approach incorporates five basic steps, which are described briefly in this section. With these descriptions, a fundamental question is directly addressed: "How would this approach benefit an economic development professional using GIS?" ArcGIS contains tools to facilitate each step of the Geographic Approach, as illustrated in figure 3-1.

These steps, in brief, can be described as follows:

1. Ask: What are the project objectives and variables?
2. Acquire: Obtain the digital geospatial data for the project and assemble the project **geodatabase**.

Five Steps of The Geographic Approach

Figure 3-1. The Geographic Approach provides a consistent structure and planning process for all GIS-based analyses.

3. Examine: Critique the acquired data, for example, by using the ArcCatalog application to generate summary statistics and preview the map features and their associated tabular attributes.
4. Analyze: Determine methodology and sequence of operations, process the data, and evaluate and interpret the results using the ArcMap application.
5. Act: Create final products for the intended audience and document methodology so that others on the team can easily duplicate the workflow and share results with others.

The next section will consider each step from an ArcGIS-centric perspective.

Step 1: Ask

Framing the research question and project objectives around a location-based perspective is important because "being as specific as possible about the question you're trying to answer will help you with the later stages of the Geographic Approach, when you're faced with deciding how to structure the analysis, which analytic methods to use, and how to present the results to the target audience."[2] Using the example from chapter 1, such a question might be, "Where are the labor pools to support the regional biotech industry?" or, more specifically, "What is the spatial distribution of research universities and colleges that grant biology degrees?"

Step 2: Acquire

Once the research question is appropriately framed, the next task—sometimes the most challenging—is to find the digital datasets needed for the research. This may entail downloading datasets via the Internet (for example, demographic data from the US Census Bureau,) using curated datasets in Esri Business Analyst Desktop (BA Desktop) or Esri Business Analyst Online (BAO) (explored in chapter 2), creating data by updating values in the tables attached to map features, or by generating new map layers in ArcGIS, such as the boundaries of the project study area. Chapter 1 discussed a variety of basic geospatial datasets obtained online from public sources, including points representing the locations of universities, polygons representing county boundaries and census tracts in the San Francisco Bay Area, and lines representing major highways. In economic development analysis, proxy measures might be warranted, such as using certain datasets as reasonable substitutes for others (for example, income distribution).

Step 3: Examine

Close examination of the data will determine if they are useful for directly addressing the research question defined in step 1. ArcCatalog, the data management and organization component of ArcGIS, is the best choice for this step as the "hub" of information in the ArcGIS environment. ArcCatalog lets users preview a dataset's geographic features and associated tabular attributes before committing to using the data to make a map. Figures 3-2 and 3-3 demonstrate how census tract map features and tabular data can be previewed in ArcCatalog (your interface may vary slightly from the ones shown in this chapter, depending on the version of ArcGIS software used).

ArcCatalog users also can determine if a dataset has a specific attribute of relevance to their project and if its values are correct. Additionally, sorting of a table's records by the values in one or more columns can be performed, and basic summary statistics describing a column's values can be generated.

While all of these tasks can also be performed in ArcMap, it's much faster to preview data in ArcCatalog since its user interface has been specifically designed for this "quick-view" functionality. ArcCatalog also allows review of the **metadata** for each dataset—information prepared by the author of the data that describes their contents,

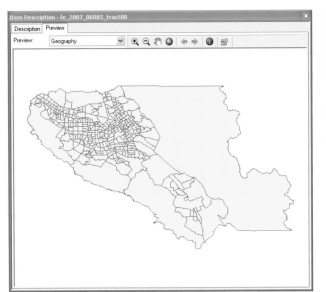

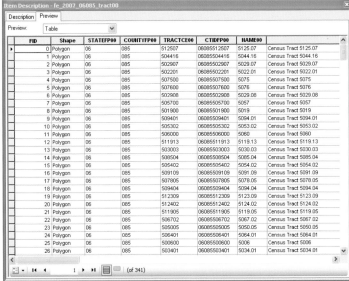

Figures 3-2 and 3-3. Previewing Santa Clara County, California, census tract information in ArcCatalog. Note the selection of the "geography" view in figure 3-2 at left image and the "table" view in figure 3-3 at right. In a GIS, each map feature (census tract polygons, in this instance) is linked to one record (row) in an attribute table. Courtesy of US Census.

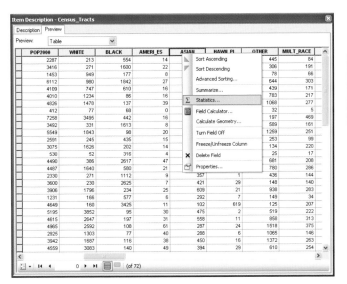

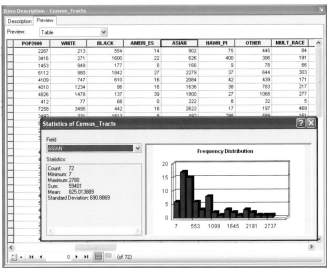

Figures 3-4 and 3-5. Generating summary statistics for the census tract dataset. Right-clicking an attribute field name (here, "Asian") and selecting "Statistics," as shown in figure 3-4, generates the summary data seen in figure 3-5, the Asian population of Santa Clara County, California, along with a histogram that illustrates the distribution of data values.

What is metadata?

The term metadata refers to a summary document providing content, quality, type, creation, and spatial information about a dataset. A memorable (though less comprehensive) definition of metadata is "data about the data." Metadata is not the same as "descriptive statistics," which is addressed in chapter 8. The data author writes the metadata and, if the author prepared it completely, it helps answer important questions, such as the following:

- Who are the data authors, and what citations should be used?
- How can the data author be contacted if questions arise?
- Can the data be freely shared, or are there restrictions placed upon its use and dissemination?
- How current are the data, and how often, if ever, are the data updated?
- What is the primary purpose of the data?
- What is the spatial extent of the data layer?

ArcCatalog contains tools that allow the metadata to be read and edited, perhaps to add notations, such as the URL used to access the data online, or notes taken during phone or e-mail conversations with the data provider.

Not all data authors take the time to prepare metadata. For example, the biotech firm data examined in chapter 1 did not arrive with metadata since the provider, BayBio, a trade association focusing on the life science industry in Northern California, creates and gathers data to promote its purposes and is purely internal, private, and not federally funded. On the other hand, all datasets created via federal funding are required to contain comprehensive metadata, using standards set by the Federal Geographic Data Committee (FGDC). All US Census data, for example, contain detailed metadata. An example of FGDC-compliant metadata is shown in the following figure. The dataset here represents the Santa Clara County census tracts, obtained from the US Census Bureau.

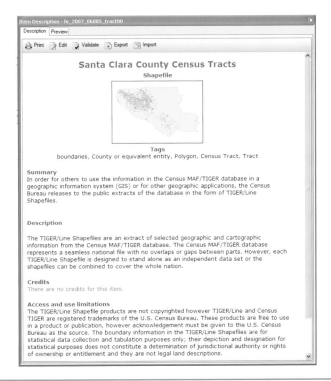

Figure 3S-1. An example of metadata for Santa Clara County census tracts, obtained from the US Census Bureau and viewable in ArcCatalog. Courtesy of US Census.

limitations and usage restrictions, if any, and other important details. (The sidebar, "What are base maps and thematic maps?" provides more information later in this chapter.)

Additionally, and to reinforce the role of ArcCatalog as a "data hub," it is the appropriate tool for copying, moving, and organizing geographic datasets, since it is designed to recognize a number of behind-the-scenes system files associated with certain GIS file types; in other words, as GIS files are moved from one folder to another, using ArcCatalog ensures that all parts of a GIS data file are moved properly. ArcCatalog is also the starting point for creating new geodatabases to organize project data in one centralized file, which this chapter will discuss more in detail.

The end result of step three of the Geographic Approach is the ability to determine conclusively if the project will include the necessary and best raw material to commence the analysis, or if the acquisition of new or better data is warranted.

Step 4: Analyze

After checking relevance of datasets to the research question by examining them in ArcCatalog, they are added to ArcMap (from a centralized geodatabase), where they become map layers that automatically overlay one another. Using the tools of ArcMap, "what if" analyses can now be conducted, more detailed summary statistics can be generated, and spatial patterns and relationships can be discovered. The results can be reviewed at any time to see if they are useful or valid. Once datasets are added to ArcMap, the data can be queried to, for example, remove or "hide" records that have missing information, or edit data values if mistakes are found or more updated information is obtained.

Recall that seven commonly used GIS tools were employed in our biotech example for the analysis in chapter 1: Table Join, Spatial Join, Geocoding, Select by Attributes, Symbolizing quantitative data, Buffer, and Intersect. This chapter will examine some of these analytical procedures in more depth by introducing ModelBuilder, an ArcGIS mechanism that facilitates the creation of executable, schematic diagrams that illustrate and document analytical approaches that a project team can consider.

Step 5: Act

The final step of the Geographic Approach determines what to do with the results. This could mean preparing and printing a map for a quarterly report or client presentation, exporting the analysis results to a website, or sharing them with colleagues via e-mail or other means. This is also the appropriate time to document methodology while it is still fresh so that the project team can easily duplicate the team leader's workflow. ArcCatalog's metadata tools come in handy at this point, allowing users to create a methodological document that can be attached to their ArcGIS maps and geodatabases.

Applying the Geographic Approach to economic development projects

Careful methodological design at the outset of a GIS project will help avoid many hours of unnecessary and redundant work, which would otherwise result from a haphazard and unstructured approach. Just about any project undertaken with GIS benefits from a deliberate, premeditated, linear approach. Adopting and agreeing upon this GIS project management approach (especially in concert with the team) will help everyone stay on track. This is true even for economic development managers who will direct others to conduct the GIS analysis.

Preparation and organization are essential at the outset of a GIS-based project to achieve desired outcomes. The work becomes easier as team leaders and manager become more adept with GIS and an organized approach takes root through repetition and adaptation. This is particularly necessary in an era when managers and their teams must do "more with less" (less *time*, in particular). Now it's time to fine-tune the Geographic Approach in a way suitable for economic development projects by referencing the Bay Area biotech analysis undertaken in chapter 1.

Determine the project objectives and variables

As with any project, the most important first step is to identify the problem, then define objectives to help break down the problem into achievable steps and measurable outcomes. At this stage, the following questions are appropriate:

- What research question forms the crux of the economic development analysis?

  Using the example from chapter 1, a research question might be "What are the factors that lead to the clustering of biotech firms in the San Francisco Bay Area?" Other questions might be "What impact does research funding in biology have on industry clustering?" and "How is industry clustering and expansion affected by the capacity of public industry-specific training programs?"

- What is the purpose of the project?

  Using the biotech example, a possible answer to this question might be "to determine if those factors fostering the clustering of biotech firms in the San Francisco Bay Area are applicable to the study region."

- What is the geographic (or spatial) extent of the study?

  This is a fundamental and necessary question to ask at the outset of any GIS project since the answer will directly affect the geographic scale of the analysis and map outputs, as well as the choice of geospatial data sources available for the project. For economic development purposes, the spatial extent might be as varied as a metropolitan region, a single county, an entire state, or a collection of census tracts or ZIP Codes. For example, in the chapter 1 biotech example, the spatial extent was the nine-county San Francisco Bay Area. Later in the example, the analysis narrowed to focus only on Santa Clara County.

- What is the basic geographic unit of analysis?

  The intent of this question is to understand the basic spatial unit into which the data will be aggregated and displayed. In the Santa Clara County portion of the biotech example, the geographic unit of analysis was census tracts; that is, the analysis was performed at that level and not at a finer-grained unit, such as census block groups or census blocks. Economic development analysis involving GIS often combines data from different sources, applies different levels of aggregation, and uses discrete location information (such as biotech firms or research university locations). This may present unique, project-specific implications for statistical analysis apart from the best practices/workflow recommendations covered in this chapter.

Identify spatial data needs and create the project geodatabase

Once the project objectives, purpose, spatial extent, and geographic unit of analysis have been determined, the focus of the GIS project shifts to data collection. As seen in chapter 2, BA Desktop and BAO provide simple access to robust datasets. Additionally, ArcMap contains a function for users to stream a wide variety of predesigned map layers (for example, roads, water features, terrain) directly into a map via the web, using an Esri-hosted service called ArcGIS Online. As long as an Internet connection is maintained, these remotely added datasets can be used right alongside locally stored datasets to create detailed maps for analytical purposes. Figures 3-6, 3-7, and 3-8 show ArcMap with an Esri-hosted base map added remotely via the web.

However, additional data relevant to the project study area will likely be needed at some point. These data, such as property boundaries from a local municipality, current zoning designations for a city, or a listing of business addresses maintained by a local chamber of commerce, are not available via Business Analyst or ArcGIS Online.

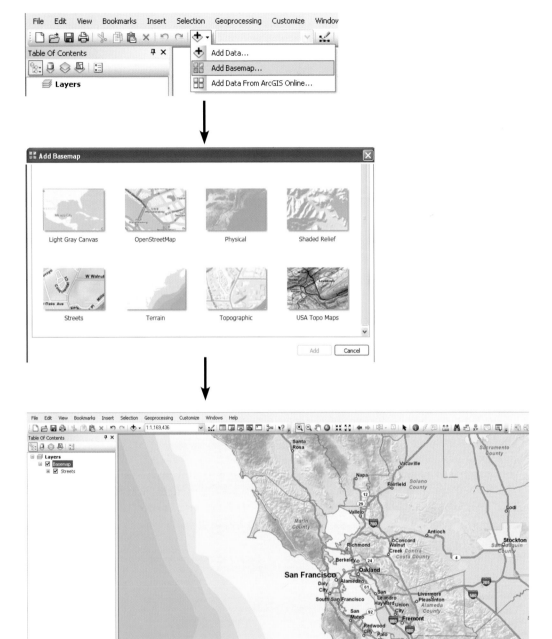

Figures 3-6 to 3-8. It is easy to add detailed, predesigned base maps to the ArcMap display via the web-based, Esri-hosted ArcGIS Online service. Here, a base map showing cities and major roadways has been streamed into the display via the web after only three clicks, as shown in figures 3-6 to 3-8. From here, users can overlay project-specific map layers. Sources for figure 3-8: Esri, DeLorme, NAVTEQ, TomTom, USGS, Intermap, Esri Japan, METI, Esri Hong Kong, Esri Thailand, Procalculo Prosis

What are base maps and thematic maps?

Once loaded into ArcMap, the layers of digital geographic data can be grouped and symbolized to form a base map (for orientation purposes, including such foundational layers as roadways, water features, and jurisdictional boundaries). Once this foundation is in place, users can add **thematic map** layers: theme, or topic-based, datasets overlaid on top of the base map.

Consider the images in this sidebar. In the first, a base map of San Francisco streets, shoreline, parks, and other orientation materials were added using shapefiles freely downloadable from the city's website. In the second, census tract boundaries (in red)—a thematic map layer containing numerous demographic attributes—has been overlaid on the base map. From here, a project team can symbolize census tract data for particular demographic variables of interest, such as race, income, and total population.

It is helpful to think of a base map as "anchoring" the project map with features so readers can orient themselves to the thematic information added on top. Without a base map, thematic layers would be "floating in space" with nothing to orient readers to specific locations.

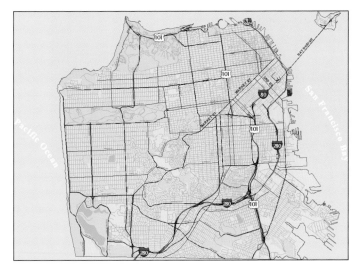

Figure 3S-2. A base map of the city of San Francisco showing basic orientation features: roads, open spaces, and water features. Courtesy of City and County of San Francisco Enterprise GIS.

Figure 3S-3 The base map of San Francisco with a thematic layer (red polygons, representing census tract boundaries) overlaid. Courtesy of City and County of San Francisco Enterprise GIS.

Since the availability and quality of geospatial data are the ultimate factors that will affect project success, determining where to find reliably accurate data is often the facet of a GIS project that takes the most time at the outset.

Fortunately, an abundance of GIS data is available for free or at low cost from myriad trusted governmental agencies and private companies, such as Esri (via Business Analyst, for example). Numerous online geospatial data portals (such as the federal government's Geo.Data.gov website, shown in figure 3-9) make the hunt for consistently reliable data easier. Additionally, significant amounts of data are available from Esri with the purchase of ArcGIS, including terrain data (useful for creating relief maps), imagery (including high-quality aerial photographs), and street networks for many countries and continents.

Figure 3-9. The federal government centralizes a great deal of digital geographic data within a geospatial "one-stop" portal at Geo.Data.gov, one of thousands of consistently reliable online data portals where GIS datasets can be acquired for use in ArcGIS.

While relatively easy to access, all digital geospatial data must nonetheless be scrutinized and selected judiciously. As a general proposition, team members of a project involving GIS should regularly review data sources that are likely to be useful. It is helpful at this stage of the project to begin listing the datasets and variables of relevance and ideas for where such data might be obtained. Table 3-1 shows what a preliminary list might look like, using the biotech example from chapter 1. All team members should endeavor to formalize this kind of clear project documentation.

Along with identifying the relevant project datasets, economic development analysts often must identify the significant *variables of relevance.* It is unlikely for all the variables to come from one source, so the project manager often must coordinate with the team to design a project geodatabase that has the desired variables and then evaluate each one in terms of its quality. Typically, the precise variable suggested by economic theory is not available, so one ends up using variables that to a greater or lesser extent are proxies. For example, data derived from individual surveys may not include highly valuable information about one's exact income or home address. In this case, the analyst might instead use a less desirable but reasonable proxy of median income data, aggregated by census tract or census block group.

Table 3-1. An example of a listing of datasets used in a GIS project, listing dataset names and sources

Dataset	Data source
Census tracts, year 2000	US Census Bureau; American FactFinder website
Census demographic data from Sample File 3	American FactFinder website
California Counties	California Spatial Information Library
California biotech firm locations	BayBio
Major biotech research universities in California	Selected by the authors from a master list of grant-receiving institutions affiliated with the National Science Foundation and National Institutes of Health
Biotech funding sources	National Science Foundation, the National Institutes of Health, California Institute for Regenerative Medicine
Degree-granting programs of relevance to the biotech industry	California Postsecondary Education Commission; nationally, similar data are available from the National Center for Education Statistics and US Department of Education
Public institutions granting BA degrees in biology	Same as above

Describe the project methodology and perform the analysis

It is worth the investment of time to construct a spatial analysis process list, as demonstrated in figure 3-10, that details the steps of the project. This attention to detail at the outset of the project will make its success much more likely and will serve as a reference during later stages of project documentation and refinement. Also, maintaining a process document is helpful after absences so that notes and ideas from long ago are at the ready. Using the biotech clustering example from chapter 1, such a spatial analysis process methodology might be listed as shown in figure 3-10.

☐ Add the California counties data set to the project Map Document

☐ Add the biotech firm address locations data set to ArcMap

☐ Zoom in to the 9-county San Francisco Bay Area

☐ Perform a count of biotech firms by county using the "Select by Location" tool

☐ Select counties with at least one biotech firm

☐ Prepare a frequency distribution graph of selected biotech firms by county

☐ Add the California research institutions data set to the project Map Document

☐ Create 5-, 10-, and 15-mile multi-ring buffers from research institutions

☐ Count the number of biotech firms within each buffer ring

☐ Add the degree-granting institutions data set to the project Map Document

☐ Zoom into Santa Clara County; add the census tracts data set to the project Map Document

☐ Create a graduated color map using census attributes that reflect the target population

☐ Select the census tracts that are underserved (far from a college to serve them)

Figure 3-10. An example of a spatial analysis process using ArcGIS; creating such lists lets project team members document workflow, which can be saved in a metadata document in ArcGIS so that colleagues can replicate the steps or edit them as project needs warrant.

Having the methodology in place creates a clear "road map" to guide the analysis of a potentially complicated GIS project. Along the way, adjustments or additions to the methodology often are warranted. When that happens, it's best to update the methodology documentation right away because it will save stress and time later in the analysis.

Once the analysis is complete, it's time to evaluate the accuracy and validity of the results to see if they seem logical. For example, in the biotech clustering example from chapter 1, if the ArcMap-generated count of biotech firms in the multi-ring buffers consistently yielded lower-than-expected totals within each ring, this might suggest a needed revision to the analysis or a repeat of the steps and tool settings to see where the problem arose. While this point might seem obvious, it is worth noting that ArcGIS cannot predict desired outcomes—the software simply executes the procedures and tools as instructed. In other words, remember never to accept the results of the GIS analysis at face value without first spot-checking the results to see if they are logical. While ArcGIS is a powerful software suite, the economic development analyst must determine what is ultimately accurate and reasonable.

An economic development analyst should always have a list of descriptive statistics for each variable used in the analysis. "Descriptive statistics" is a generic term, but it usually consists of a variety of measures concerning individual values (attribute fields), such as mean, median, mode, standard deviation and—perhaps most important—the minimum and maximum values of the variable being examined. This verification tool helps analysts identify problems at a glance, such as an income figure that appears as a negative number or a value in the results that is impossibly high. Figure 3-11 shows what such statistics might include.

	A	B
1	*FTE Graduate Enrollment*	
2		
3	Mean	1219.273913
4	Standard Error	211.0359039
5	Median	1136.6
6	Mode	842.677
7	Standard Deviation	1012.09264
8	Sample Variance	1024331.513
9	Kurtosis	-0.901252054
10	Skewness	0.655625495
11	Range	3144
12	Minimum	0
13	Maximum	3144
14	Sum	28043.3
15	Count	23

Figure 3-11. An example of descriptive statistics for full-time equivalent graduate enrollment at the twenty-three campuses of the California State University system. Such statistics provide a highly useful snapshot of the dataset and are a component in verifying data accuracy.

Descriptive statistics, including those generated by ArcGIS, are discussed further in chapter 8, along with a more systematic approach to "reality checks" involving statistical data. There are a few ways to generate data summaries in ArcGIS. For example, as seen in figures 3-4 and 3-5, from within data tables a project team can choose an attribute field (for example, a column containing data values that correspond to map features) and quickly generate summary statistics by right-clicking the relevant attribute column heading. Basic summary statistics for the selected attribute include minimum value, maximum value, sum, mean, and so on (in figure 3-12, the total population of San Francisco census tracts, per the US Census Bureau), which are displayed along with a histogram of the data values.

Another commonly used data summary tool is the Summarize feature, also accessed by selecting an attribute heading in the data table, as shown in figures 3-13 and 3-14 (your interface may vary slightly from the ones shown in these figures, depending on the version of ArcGIS used).

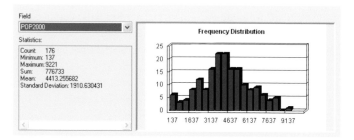

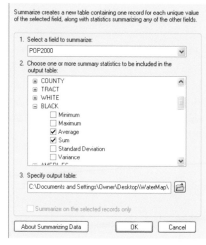

Figure 3-12. An example of summary statistics produced by ArcMap. Here, the population within San Francisco census tracts is summarized. Note the histogram of data value distribution at the right and the total population (776,733 in the year 2000).

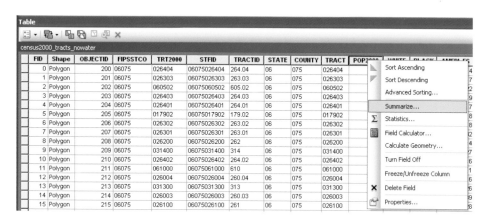

Figures 3-13 and 3-14. Accessing the Summarize function in ArcMap begins by right-clicking the attribute column of interest, as shown in figure 3-13 in which the total year 2000 population attribute has been selected for summarization. Figure 3-14 shows the Summarize dialog that opens. The dialog contains three steps: the first step displays the selected attribute field automatically, though it can be changed. The second step allows for an additional attribute to be summarized along with the attribute from the first step; here, the total and average African American population per San Francisco census tract will be calculated. The location for the summary output table is chosen in the third step.

Present the results

Once the analysis results are deemed accurate, they can be presented in a suitable format for the intended audience. This could take the form, for example, of a large printed map for display in a boardroom, a digital map for a web page, or a page in an economic development report. ArcGIS offers many methods to export map results to numerous print and digital media formats, such as the common PDF format. Additionally, ArcMap offers a number of wizards that take users through a series of steps for quickly and easily preparing project data reports, complete with predesigned, professional-grade templates.

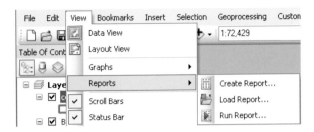

Figure 3-15. ArcMap contains report-generating capabilities that walk users through the process of producing professional-grade summary reports.

A GIS project will have more than just one potential output (and more than just one format for any given output). For example, in a team-based project, the results generated by one member might serve as the input for another team member's portion of the analysis that may or may not use GIS. Consider the example of an econometric study: a GIS analyst might perform spatial analysis that results in numeric results, such as summary statistics for income within census tracts. These outputs might then be used in statistical analysis software to generate econometric results and these, in turn, might be returned to the GIS analyst to create maps that illustrate the findings.

Document the work and share findings with the project team

Documenting the project's methodology and summarizing the data analysis and study conclusions will let the project manager and others duplicate the workflow if needed. The information can be entered into ArcCatalog or prepared in word processing software, then imported using ArcCatalog into a summary document, attached to the project map, and accessed by anyone on the project team.

The final step involves preparing the maps and data for sharing with the project team. As GIS technology blossomed in the 1980s and early 1990s, most practitioners tended to work on stand-alone computers to conduct their analysis. If data and maps were shared at all, the process was often cumbersome, involving the swapping of disks and other limitations forced by the technological limitations of the era. The Internet made the sharing of information increasingly easier as FTP sites, e-mail, and data sharing websites took root. It is relatively easy to share maps and data with colleagues via an ArcGIS capability called **map packages**. Map packages contain the map document (the file that captures the map design settings, symbols, and colors) and the datasets that constitute the map. Each document is packaged into one portable file. Map packages can be used for easy sharing of maps between colleagues in a work group, across departments in an organization, and with ArcGIS users via ArcGIS Online, www.arcgis.com, a web-based collaboration tool (figure 3-16).

Figure 3-16. ArcGIS Online makes it easy to share maps and data with a project team. This image shows the My Content interface (your interface may vary slightly, depending on the version of the website in operation). The Groups tab at the top lets users create specific workgroups for sharing project maps and data.

Map packages have other uses, too, such as the ability to create an archive of a particular map that contains a snapshot of the current state of the data used in the map.

The GIS project workflow presented above represents a time-tested approach to preparing for and conducting analyses using ArcGIS, founded on the structure and mindset of the Geographic Approach. For economic development projects, this structure brings the necessary order to the process of individual and collaborative geospatial analysis. Furthermore, when this process is coupled with the ease by which data and maps can be shared using the web and tools such as ArcGIS Online, the basic elements are in place to coordinate all facets of a GIS-based project.

Now the focus can turn to the all-important "fuel" of GIS-based projects—the data. This section begins with an overview of matters related to data quality assurance and continues with a discussion of the two primary data formats to represent geographic features on digital maps—vector and raster.

Ensuring data quality

The first two chapters showed how a number of geospatial datasets and a handful of ArcGIS tools could be used in concert to complete a basic economic development analysis. As the examples demonstrated, geographic features representing real-world entities as points, lines, and polygons (and their associated tabular attributes) form the building blocks of a GIS map. As such, the *quality* of digital data used in GIS-based analyses is essential to the validity of the analyses. The procedures employed to obtain, inspect, manage, and share the data are essential to the success of the project.

Just as a master chef learns as much as possible about the ingredients that comprise a dish, so must those involved in a GIS project carefully consider the geospatial data available for an economic development analysis. To extend the analogy, the quality and accuracy of the data—the ingredients of a map—are fundamental to producing meaningful results in a GIS and understanding the reputation and practices of the providers of geospatial data.

Many people are familiar with GIGO, the acronym for "garbage in, garbage out." In a GIS, the quality of the inputs to a process has a direct bearing on the quality and reliability of the output. For example, inadvertently using incomplete, inaccurate, or outdated data in a map can yield erroneous or misleading results. Conversely, starting with data that have been evaluated for quality, completeness, and timeliness means the GIS user can more confidently stand by the findings generated in an analysis. In other words, know the data (map features, tabular attributes, and metadata) before starting. As discussed earlier, ArcCatalog is the tool of choice for this task since its user interface is specifically designed for previewing, inspecting, and evaluating the attributes and metadata related to the project datasets.

Representing geographic features digitally with vector and raster data formats

Remember that all geospatial data and maps are graphic representations of reality; therefore, some simplification of features is inevitable and necessary. So, when it comes to the task of representing real-world features digitally, two primary formats—vector and raster—are available. Understanding both is a GIS best practice because a project may require one or the other (or both), or only one format may be available from a data provider.

Vector data represent discrete, clearly delineated features on maps. Geographic entities in the vector format are classified into three categories: points, lines, and polygons. These are illustrated in figure 3-17, a map comprised entirely of vector data representing coffeehouse locations in San Francisco. Note the use of vector lines representing roads, polygon vectors representing parks (in green) and water features (in blue), and vector points representing the locations of coffeehouses (in red). With the exception of the coffeehouse data, which was created by the authors using data from coffee company websites, the information was acquired for free from the city's enterprise GIS website.

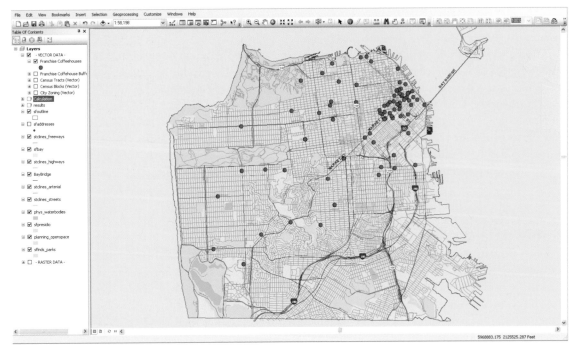

Figure 3-17. Locations of coffeehouses in San Francisco, using vector data in all three of its forms: points (red dots representing coffeehouse locations), lines (representing roads), and polygons (representing parks, in green, and water features, in blue). The roads, parks, and water features together form a base map for orientation purposes, upon which the coffeehouse thematic dataset is overlaid. Courtesy of City and County of San Francisco Enterprise GIS.

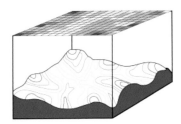

Figure 3-18. Raster data depict elevation information, in the lower section, and how this might be represented in the raster model, at the top. In this illustration, the designer has chosen to depict elevation peaks with darker colors in the raster model.

Whereas vector data are useful for mapping *discrete* geographic features, raster data are used to represent features that are *continuous*; that is, those that lack clear, distinct boundaries and whose intensity and distribution vary across a given area. Three examples of continuous phenomena are elevation, precipitation, and temperature. Consider figure 3-18, showing elevation contours in a hilly area depicted at the bottom of the image.

Note how this undulating terrain can be depicted in the raster format, represented here by an array of cells (top part of the image), with each cell assigned a numerical value corresponding to a particular elevation at a given ground-level point. The darkest green cells in this raster correspond to the elevation peaks, and the lightest green cells reflect lower elevations.

Another example of raster data is shown in figure 3-19, with land-cover data collected by NASA satellites, which is particularly useful for determining areas of urbanization, forest cover, water, and other categories. NASA has color-coded the dataset cells to differentiate land cover in San Francisco. When overlaid with vector data for

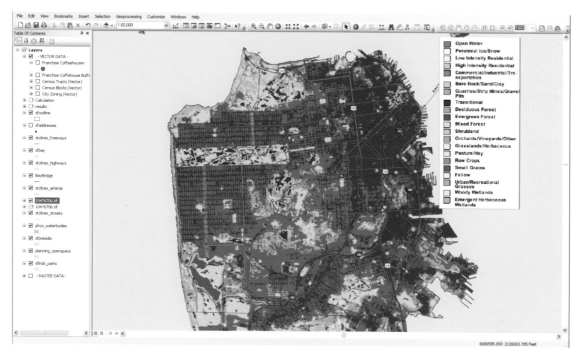

Figure 3-19. Land-cover raster data collected by NASA satellites, symbolized (colored) by major land-cover category, as reflected in the legend. Each pixel in this image represents thirty square meters on the ground. The predominance of red on this map corresponds to urban land uses. Land-cover data are a particularly useful input for certain economic development projects, such as site-suitability analyses. Courtesy of City and County of San Francisco Enterprise GIS.

orientation purposes (perhaps by adding highways or parcels of property) the value of GIS to display vector and raster data simultaneously becomes evident for many purposes, such as site selection.

Raster datasets are often used to provide backdrops to maps, such as aerial photographs and, as shown above, satellite-derived data. The simple structure of raster data—an array of evenly spaced cells, each tagged with a numeric value—belies the sophisticated mathematical analysis that can be conducted using this geospatial data format, especially when multiple, overlaid raster datasets are used in concert to conduct, say, site-suitability analyses. The use of raster data for economic development analysis will be more closely examined in chapter 9.

While neither the vector nor the raster model is inherently superior (each has its pros and cons in terms of file size and precision, data availability, and relevance to particular projects), the GIS project manager and support staff nonetheless need to be conversant with both formats to determine the most appropriate data types needed for an economic development analysis. For example, in a site-suitability study for a proposed distribution center, the GIS team might incorporate both types. Raster data can represent the steepness of slopes to find areas that are flattest and therefore require minimal site grading and lower construction costs. Vector point data can be used to determine the distances between the distribution center and the businesses it serves. Other vector datasets, such as city zoning and census demographics, can be added to the mix.

Using geodatabases for organized management and storage of GIS data

Using a file geodatabase is the best way to organize the numerous datasets involved in GIS-based projects before taking the next step—creating a map.

Developed by Esri, file geodatabases are built on the relational database model, in which multiple datasets and their corresponding tabular attributes are all contained within one file. This process allows relationships to be established based on shared attributes. File databases match data elements by using common characteristics found within the dataset and organizing these elements into groups. For example, a file geodatabase containing vector point locations of customer locations can be grouped by industry sector or the customer's name. Think of file geodatabases as "containers" designed to hold a collection of various types of GIS-based datasets, including vector data, raster data, spreadsheets, transportation networks, and more. The icon used to represent a geodatabase in ArcGIS looks like a cylindrical container, reinforcing this "catch-all container" trait.

Figure 3-20. Two file geodatabases, one titled "Raster" and the other "Vector," as displayed in ArcCatalog. The "Vector" geodatabase has been opened to reveal a number of feature datasets inside, each containing specific vector datasets that can be added to a project in ArcMap. The folder titled GIS_CCA contains the entire set of data and map documents for an economic development project; this attention to careful data organization will pay off in time savings and team collaboration as a GIS-based project evolves.

Figure 3-21. The ArcCatalog dialog box for importing vector, raster, or tabular data into a file geodatabase. Commonly used shapefiles, a vector file type, can be imported into the geodatabase, after which they become known as feature classes. Organizing project data into one file geodatabase is advantageous for data storage and team collaboration.

File geodatabases offer many data management and collaborative benefits, in part because each one can hold all the project data in one file (with map documents contained in a separate but directly linked folder). Additionally, file geodatabases are useful for a number of other reasons:

- They are easy to share with anyone because they can be e-mailed, copied to an FTP site, or copied to a disk.
- Because file geodatabases have no effective size limit, they can hold vast amounts of information (up to terrabytes!) and still operate correctly and efficiently.
- Preset coded values can be set within geodatabases to establish validation procedures—that is, a list of acceptable values that can be entered into attribute fields via drop-down lists—as opposed to open-ended text entry fields. This helps to avoid data entry mistakes.
- Specific elements inside file geodatabases, such as data tables, can be exported to other file formats, such as dBase Table File Format (DBF) that can in turn be imported into database or spreadsheet programs used commonly by economic development analysts, such as Microsoft Access and OpenOffice Base.
- Likewise, vector data inside file geodatabases can be easily exported to the Esri shapefile format using ArcCatalog, which some users prefer because the US Census Bureau, for example, provides a great deal of GIS data through this format.

Automating GIS analysis with geoprocessing

The term geoprocessing might initially seem intimidating, but breaking the term down into its constituent parts, it is *geo* and *process*: that is, a process applied to geographic data. Any process undertaken involves three things: inputs, a process of some sort, and resulting outputs. For example, cooking a meal requires inputs (the food ingredients), a process (mixing, cooking, and serving) and outputs (the finished meal). Likewise, a geoprocess has inputs (digital geographic datasets), processes (ArcGIS offers hundreds of tools in ArcToolbox to execute these), and outputs (new geographic datasets that result from use of the tools).

Economic development analyses that incorporate GIS require using a number of tools on a consistent basis. Chapter 1 touched upon seven of the hundreds of tools bundled in ArcGIS and organized in ArcToolbox. These tools are known collectively as geoprocessing tools. The fundamental purpose of geoprocessing is to allow one to automate GIS tasks. The manner in which these tools are accessed and linked together to form analytical models is the focus of this section.

The tools organized within the ArcToolbox application provide functionality to serve the analytical needs of a wide range of disciplines. Once opened, the ArcToolbox window reveals a tree-view user interface that organizes all of the geoprocessing tools, as shown on next page.

The kinds of GIS tasks to be automated via geoprocessing can be rather mundane—for example, converting shapefiles to feature classes in a centralized file geodatabase for a project. Or, the tasks can be rather creative, using a sequence of operations to model and analyze complex spatial relationships—for example, calculating optimum paths for shipment of goods through a transportation network, analyzing and finding patterns in business locations, or predicting which areas possess the demographic and land-use attributes best suited for targeted redevelopment strategies. Since economic development analyses are inherently multivariate and sequential in nature, geoprocessing is likely to feel quickly intuitive to economic development practitioners.

Geoprocessing involves running prewritten computer programs, though it rarely feels like programming, given the seamless way in which it is handled in ArcGIS. One example of a geoprocessing operation is to use the Multi-ring Buffer tool that was employed in the chapter 1 example and illustrated in figure 3-23.

In BAO, all of the geoprocessing (as in the generation of drive-time distances demonstrated in chapter 2) occurs behind the scenes, whereas in ArcGIS users can control and customize geoprocesses to suit their economic development project needs.

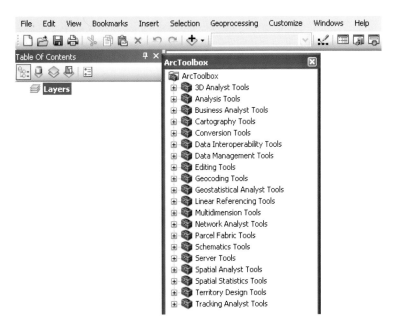

Figure 3-22. The ArcToolbox window opened in ArcMap. Inside of the toolsets shown here (such as ArcGIS 3D Analyst extension tools, Business Analyst tools, and so forth) are individual, executable geoprocessing tools such as Buffer, Intersect, and hundreds more.

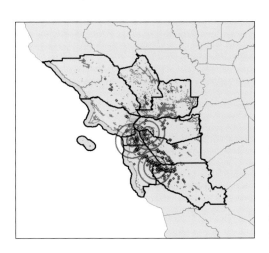

Figure 3-23. Recall in chapter 1 that the Multi-ring Buffer geoprocessing tool was used to measure and display five-, ten-, and fifteen-mile distances (red concentric rings) from each Bay Area research university as part of an econometric analysis focused on access to skilled labor. The resulting multi-ring buffers became a new map layer. Courtesy of US Census; BayBio.

The choice of "ease of use" versus "full control" of geoprocessing tools ultimately depends on the scope and complexity of the economic development project. For quickly generating results and reports, for example, it might be sufficient to run a simple geoprocess using BAO's easy-to-use tools. Other cases might require the ArcGIS geoprocessing environment for customizing tools and settings to better support your project needs.

The manner in which geoprocessing tools are accessed can be explained by way of an analogy to spreadsheet programs. Like Esri's GIS tools, spreadsheet programs typically offer prewritten, built-in functions to accomplish specific tasks. For example, consider one can invoke a function (say, the computation of the natural logarithm) that, when executed, runs a program "behind the scenes" to yield desired results. In other cases, however, one might wish to "record a macro" to document a process. In a spreadsheet program, a macro can record mouse clicks and keystrokes and let users play them back later. This process can record a sequence of commands relevant to the task at hand. When the macro is executed, it plays those exact commands back in the same order; the spreadsheet program behaves as if the user had entered the commands.

Recording a macro documents a workflow: a series of steps to achieve desired results. Likewise, geoprocessing in ArcGIS supports the automation of workflows by providing many tools as well as a mechanism to combine the tools together in a sequence of operations. The ArcGIS mechanism for this is called ModelBuilder.

Using ModelBuilder for geoprocessing

Each geoprocessing tool performs a small yet essential operation on geographic data, such as adding a new field to a table or creating a buffer zone around features to define drive-time distances from store locations. ArcGIS includes many such geoprocessing tools that assist with simple tasks such as this, as well as more detailed tasks including three-dimensional (3D) surface analysis, geocoding of address data, transportation and goods movement studies, and geostatistical analysis, including "hot spot" identification (finding areas of statistically significant concentration such as customers or store locations). The geoprocessing workflow can be represented with a simple diagram as shown in figure 3-24.

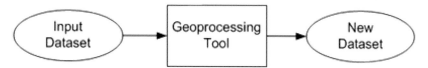

Figure 3-24. The basic conceptual geoprocessing workflow representing the elements needed to form a geoprocessing model in ModelBuilder: input datasets, connectors (arrows), geoprocessing tools, and output (new) datasets that result from using the tool.

Geoprocessing allows the ArcGIS user to chain together *sequences* of tools, using the output of one tool to serve as input for the next. This sequencing ability can be used to compose powerful geoprocessing models (or tool sequences) that help automate multistep project work and solve complex problems.

Models visually represent geoprocessing workflows in ArcGIS. This opens the door for opportunities to preserve and share workflows. ModelBuilder visually depicts and then runs a sequence of model steps in an analysis. Users set up a schematic diagram of linked input datasets, tools for processing operations, and output datasets, much like a flow chart, with no programming required. Many people become adept at GIS by using ModelBuilder since it facilitates the design and manipulation of an intuitive diagram that helps to visualize, for example, a multistep economic development analysis involving GIS, such as the one presented in chapter 1's study of the spatial relationship between population centers, research institutions, and biotech firms.

Exploring ModelBuilder

The schematic diagrams created in ModelBuilder are more than just a static representation of a workflow—the diagram itself is "live" and performs the functions it depicts. That is because the diagram parts (boxes, arrows, and so on) represent mini-programs that perform the operations illustrated. For example, a project team might be interested in knowing what parts of San Francisco are located within 500 feet of coffeehouses, perhaps as a component of a marketing study. Figure 3-25 shows a basic model in ModelBuilder designed to calculate buffers around coffeehouses in San Francisco. The first two model elements—a coffeehouse location dataset and the Buffer geoprocessing tool—were added to the model window from ArcCatalog, the data management hub. The dataset and the tool were literally dragged into the model from the ArcCatalog window. Note the similarity of this model to the conceptual model in figure 3-24; here, the coffeehouse dataset is the input, the Buffer geoprocessing tool is indicated with a square, and the resulting, new buffer dataset is represented as "output feature class" (the default name).

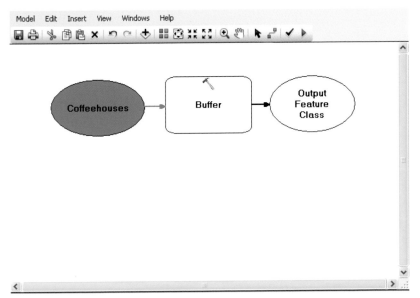

Figure 3-25. The ModelBuilder window with an element representing the San Francisco coffeehouses dataset, the Buffer geoprocessing tool, and the output dataset that will be created once the model is executed. The tool and output are not colored, indicating that user input is needed; in this case the user needs to enter the desired buffer distance.

Double-clicking the Buffer model element reveals its dialog box, where the desired buffer distance is entered along with other parameters, such as the file path to the output feature class that will reside in a file geodatabase, as shown in figure 3-26.

Figure 3-26. The dialog box for the Buffer geoprocessing tool shown in figure 3-25. Note that a desired buffer distance (500 feet, in this case) can be entered here.

Once the parameters for the Buffer tool are entered, the entire model is displayed as colored in, meaning that it is ready to run.

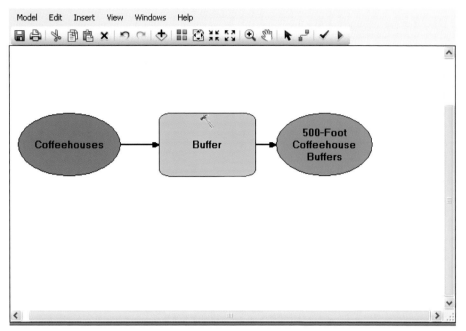

Figure 3-27. The model is now in a ready-to-run state since all model elements are colored in. Clicking the Run button (blue arrow toward the upper right of the model window) executes the model. The result is shown in figure 3-28.

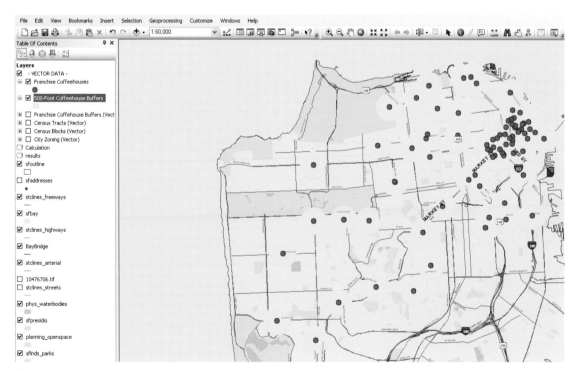

Figure 3-28. The results of executing the 500-foot coffeehouse buffer model. Note the yellow circular areas representing the buffers generated by the model as a new map feature. These buffered areas could represent the areas of San Francisco within an easy walking distance to coffeehouses. Courtesy of City and County of San Francisco Enterprise GIS.

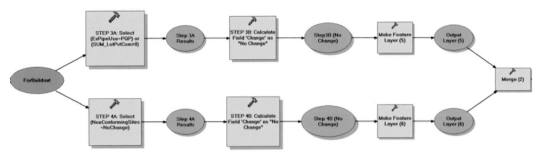

Finding all "No Change" Parcels and Eliminating them from Consideration

Figure 3-29. This model was created to accept input data (property records and blue input element) for an entire community. Through a series of iterative, linked geoprocesses (represented by the series of yellow model elements), parcels were eliminated from consideration for community redevelopment efforts.

Figure 3-29 is an example of outputs of one geoprocess in a model serving as inputs to subsequent geoprocesses. It was designed by the authors, using ModelBuilder to analyze all properties in a San Francisco Bay Area city to identify those unsuitable for the city's redevelopment proposals. The specific text shown inside of the model components is not as important for this discussion as the ModelBuilder components themselves. The blue oval represents input data (a dataset of all land parcels in the city), the yellow squares represent specific geoprocesses (such as selecting subsets of parcels based on specific criteria, including lot size and zoning), and the green ovals represent interim outputs from each geoprocess. The arrows show how datasets, processes, and outputs are sequenced and chained together, culminating in the merging of the two derived datasets, using the Merge tool shown at the far right.

The advantages of using ModelBuilder for economic development analyses

ModelBuilder is relatively easy to learn, which belies its ability to schematically document and execute complex workflows. For "quick and dirty" geoprocessing operations or for one-time, temporary analyses, ModelBuilder might not be needed, in which case the relevant geoprocessing tools can be executed directly and individually from ArcToolbox without the use of a model. However, since economic development analyses are typically multivariate and complex in nature, ModelBuilder's many advantages include:

- The ability to document, automate and preserve geoprocessing workflows

  Models often have several parallel or interdependent processes (as there are in most economic development analyses). Thinking practically, many economic development analysts must produce regular monthly or quarterly reports. ModelBuilder can make the process of updating these reports much easier by preserving the workflows. Also, the "drag and drop" functionality in ModelBuilder makes it easy to update an existing model. For example, users might create a complex model for a one-time study. Then, as often happens, they might be asked to modify the model based on the acquisition of updated information. Rather than rebuilding the model from scratch, users can just replace the older data elements with the new ones via dragging and dropping the newer data from ArcCatalog and rerunning the model. Creating models saves time and uses staff resources more efficiently, since common analytical processes do not have to be repeatedly created. Furthermore, ModelBuilder contains many features that allow documentation of work in the form of reports and graphics. For example, the Model Report function automatically produces a listing of project variables, geoprocesses and results—a major time-saver and an efficient standardization tool. Additionally, graphics of the model can be printed out and added to reports, with colleagues using its intuitive schematic design.

- Ease of use

  With its drag and drop functionality, spare and simple user interface, and intuitive tools, ModelBuilder is relatively easy to learn. A model is quite literally a program, although the designer does not need to possess any programming knowledge, only the ability to think about the logical sequence of operations and datasets related to the project at hand. With a little extra effort, iteration and looping can be built into a model in order to rapidly cycle through an ordered series of inputs. ModelBuilder is an important component of ArcGIS because it both facilitates communication among teammates working on a project and results in the steps being carried out. Even team members who have no familiarity with ArcGIS can learn to diagram problems in the way suggested by ModelBuilder—even without a computer. The time spent reading this chapter would be enough for someone without any background in GIS to communicate effectively about the steps in a complex GIS project.

- Reduce duplication of effort when working on projects

  The consistent use of ModelBuilder in a project team allows participants to take advantage of common experiences and avoid mistakes since models are easily documented, can be quickly tested, and are simple to distribute. Limited project time and financial resources can be used more efficiently. Also, by preserving geoprocessing workflows common to economic development analyses (such as the production of quarterly reports) in the form of a model, duplication of effort is significantly reduced.

- Models and data can be shared among project team members

  Models can be stored within file geodatabases along with project data. Recall that a file geodatabase is a container to hold multiple file types: vector and raster data, tables, and tools. Recall that the entirety of an economic development project's datasets, tools, and models can reside inside of a single file, which can be easily sent to a colleague via e-mail, FTP, or other means, along with any related map documents that reference the datasets in the geodatabase.

- The ability to easily conduct "what if" scenarios

  One of the primary advantages of ModelBuilder is its ability to visually document and preserve workflows as models and to quickly modify parameters to re-execute the model; this is perfect for iterative, "what if?" analyses so common to economic development studies. ModelBuilder offers the ArcGIS user the ability to quickly modify input parameters to model processes, generate new results, compare them to previous results, and determine which scenario is most suitable.

- Models are easy to update

  Envision a scenario where 2000 US Census data are used in a model to incorporate demographic variables into the study. Upon release of 2010 Census data, old data can be swapped with the new in the model and then rerun to yield updated results. This ability to easily update models with new information—including updated customer address databases, demographic data, and road networks—reinforces the time-saving nature of ModelBuilder.

- Models easily exploit data scalability

  Related to the previous point, models can be scaled upward as the project team acquires larger datasets for the project, such as new housing developments, a wider set of cities captured within the **trade area**, and larger sets of customer addresses.

Summary

The description of the GIS best practices presented in this chapter revealed the mindset and structured approach needed for successful economic development analysis using GIS. The discussion began with the Geographic Approach, an organized, sequential, logical five-step mindset that creates the overarching method for analysis with GIS. This approach was illustrated by promoting project-level management techniques for efficient GIS-based economic development analysis. The discussion then turned to data quality and an overview of the two primary digital geospatial data formats, vector and raster, and the use of file geodatabases to store project data in a single, centralized file. Finally, the chapter covered geoprocessing and ModelBuilder; the former is a term for a data processing workflow, which is foundational to almost all work conducted in GIS; the latter is a tool for structuring, executing, sharing, and saving a workflow in an intuitive, executable, sharable, flow-chart format.

Notes

[1] Jack Dangermond, "A Framework for Understanding, Managing, and Improving Our World: GIS—The Geographic Approach." *ArcNews Online,* fall 2007, accessed January 15, 2012, http://www.esri.com/news/arcnews/fall07articles/gis-the-geographic-approach.html.

[2] Ibid.

Part II

Applying the Geographic Approach and GIS to economic development analysis

4

Site selection using Esri Business Analyst: Locating a retail store in the Tampa Bay Area

Randy Deshazo, contributor

Objectives

- To illustrate the use of Esri Business Analyst Desktop in site selection
- To discuss the use of the Huff Model in connection with Esri Business Analyst
- To discuss the fiscal implications for jurisdictions of site selection

This chapter and the next two chapters are applications of GIS to specific problems in economic development. Each is organized around the Geographic Approach, as outlined in chapter 3. These are the elements of the Geographic Approach: Ask, acquire, examine, analyze, and act.

The Geographic Approach: Ask

This chapter addresses the challenge of selecting a site for a retail outlet. Site selection is one of the core activities of economic development officials. Economic development professionals want to promote their area as an appropriate place for a variety of economic activities: retail, manufacturing, and service provision. Locating retail sites is an art and a science. GIS can strongly assist with the science side, but nothing can replace the judgment and wisdom of the economic development practitioner. This chapter approaches site selection using an economic model—the **Huff Model** of retail location—and other tools in Business Analyst Desktop (BA Desktop). Chapter 9 explores ArcGIS software tools for site suitability approached in a different way, using raster data.

Although economic development professionals may be motivated to generate higher incomes, tax revenues, and more jobs in the local area, proposed projects must appeal to prospective firms and investors based on hard economic data and analysis. Often, the success of an economic development initiative depends on the availability of suitable sites for the proposed activity.

This chapter will consider the location demands of a hypothetical new "formula" retail store. Formula retailers, otherwise known as chain stores, conduct business in many locations according to a centralized management philosophy and oversight and often are distributed around the country on a franchise basis. Modeled as an amalgamation of some of the most successful retailers in the United States, this hypothetical firm, Shopping Cart, will sell moderately priced upscale retail goods and services to customers in a test market in which the firm's investors believe that the demographic characteristics of the population best reflect the nationwide target customer base.

There are several steps in identifying the best location for the store. This chapter reviews these steps and illustrates tools that BA Desktop provides to support the analysis. The chapter concludes with a section discussing the implications of site selection for municipalities' economic development and fiscal strategy.

The main considerations in locating a retail store are as follows:

- Define the study area
- Prospect for customers
- Identify competitors
- Estimate trade areas
- Examine cannibalization

Each of these elements can be addressed more surely and more quickly by using BA Desktop than by using traditional approaches to site selection. Furthermore, the Huff Model can be used to forecast sales at alternative prospective sites.

Defining the study area

The hypothetical chain retailer, Shopping Cart, which appeals to single urban professionals and upper-middle class households, has a presence in many large urban markets in the United States and Canada but has not yet placed a store in any Florida market.

While there are several large market areas in Florida, such as Miami-Dade, Orlando, and Jacksonville, this analysis will focus on entering the Florida market in the Tampa Bay Area because several potential sites have recently become available, and zoning regulations for those sites permit the proposed use of an upscale shopping store.

Currently home to over 2.7 million residents, the Tampa Bay Area is the nineteenth largest metropolitan area in the United States and the second largest in Florida after Miami-Dade. The region consists of four counties surrounding Tampa Bay: Hillsborough, Pasco, Pinellas, and Manatee. Facing the Gulf of Mexico, with some of the finest beaches in the United States, the Tampa Bay Area attracts a wide spectrum of residents—from the so-called "winter birds" (retired people who come to spend the pleasant winter months in the region) to a diverse native-born population comprising most major ethnic and demographic clusters of the American population.

Tampa Bay is also diverse in terms of housing choices, household income, and spending habits. Employment is spread among many different sectors, with concentrations in finance and defense industries. As such, Tampa Bay is a microcosm of the United States, and it is a frequent test market for formula retail chains and restaurants.

Initially, the analysis will view the problem from the perspective of hypothetical investors in a Shopping Cart franchise. The focus will be on considerations that would make the site most profitable for Shopping Cart. Because economic development officials are also keenly interested in business success, they strive to find locations where businesses are likely to be profitable. However, economic development officials must weigh additional considerations, such as increased congestion or pollution and the tax impacts in assessing the suitability of a site. Their perspective is discussed in greater detail at the end of this chapter.

Figures 4-1 and 4-2 orient readers to the Tampa Bay Area. Figure 4-3 is a map depicting areas of the highest average household income. This map is the basis for considering where a Shopping Cart store might be placed.

The Geographic Approach: Acquire

Business Analyst makes the transition from "ask" to "acquire" in the Geographic Approach quick and painless. Using Business Analyst's comprehensive database of socioeconomic characteristics of the population in the study area and spending patterns of residents, analysts can better understand the targeted market through thematic (theme- or topic-based) mapping. Figure 4-3, for example, depicts the distribution of household incomes based on US **census block groups** throughout the Tampa Bay Area. Census block groups are usually the lowest level of aggregation at which census data are readily available. Census tracts consist generally of four or five census block groups. A general rule of statistics is that aggregation destroys information; more aggregation destroys more information;

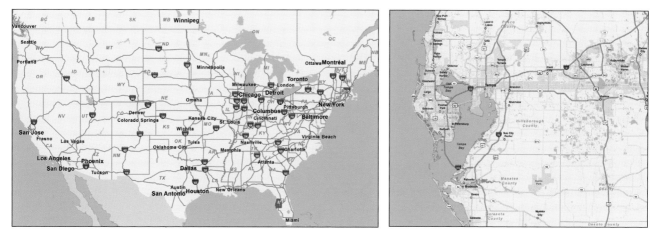

Figures 4-1 and 4-2. Location map for Tampa metropolitan area, at left, and a closer look at the Tampa area, at right. Data displayed in screenshots of Esri Business Analyst are courtesy of Esri; US Census Bureau; Infogroup; Bureau of Labor Statistics; Applied Geographic Solutions, Inc.; Directory of Major Malls, Inc.; GfK MRI; and Market Planning Solutions, Inc.

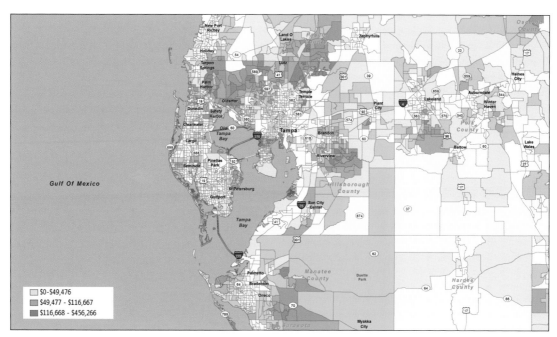

Figure 4-3. Distribution of median household income by census block groups in the Tampa Bay Area in 2010. The map shows average household income at a fairly refined level—the census block group. The darker green color corresponds to a higher average income. Data displayed in screenshots of Esri Business Analyst are courtesy of Esri; US Census Bureau; Infogroup; Bureau of Labor Statistics; Applied Geographic Solutions, Inc.; Directory of Major Malls, Inc.; GfK MRI; and Market Planning Solutions, Inc.

hence, the lowest level of aggregation is usually desirable for *analysis* but not always for *presentation* because the level of detail can overwhelm people's ability to comprehend.

As indicated in figure 4-3, there are concentrations of high household income along the coast in Pinellas County, but most concentrations of high household income are directly north of Tampa. The analysis will concentrate on locations north of the city because Shopping Cart sells upscale consumer goods.

The Geographic Approach: Examine

Business Analyst also makes examining the data straightforward. The data in Business Analyst are fully documented in metadata. The Business Analyst data arise from various sources, but the quality of the data is easily determined. Additionally, Business Analyst offers forecasts from Esri for a variety of demographic and economic variables.

The Geographic Approach: Analyze

The analysis will want to identify customers and their locations. Business Analyst includes aggregated sales data for dozens of consumer goods—from women's apparel and restaurant expenditures to school supplies and many other categories of spending—based on where the shoppers live.

Customer prospecting

If the proposed store were to concentrate on just one narrow customer category (for example, women's apparel), the analysis could simply map potential locations for the highest expenditures of those goods. The color-differentiated choropleth map in figure 4-4 depicts the distribution of highest spending levels for women's apparel. The map is in the form of a classic **heat map**—the higher levels of spending are "hot"—associated with deeper shades of red. The lower levels of spending are "cool"—designated by deeper shades of green. The dollar amounts are total spending by census tract on women's apparel in 2010.

Not surprisingly, areas with the highest median household income are also among the areas with the highest levels of spending on women's apparel. Again, our attention is drawn to the area north of Tampa. To better distinguish the areas most suitable for a Shopping Cart store, the symbology can be changed to highlight only the census tracts (green) that have the highest spending on women's apparel, as shown in figure 4-5.

Let us assume that after looking at a wide range of business and socioeconomic data, Shopping Cart decides to place a store in a particular Tampa Bay Area neighborhood, or subarea. For retail, that site should be one that balances the distance traveled by the most potential customers with the trade area penetration of the store.

Trade area penetration is the share this store will have among all other competitors of all potential customers in the region. Competitors are defined as those who sell the same or similar products. The number of competitors will vary, depending on what constitutes "same or similar" product, so the trade penetration measure will vary. For example, Shopping Cart's competitors certainly include Macy's and Nordstrom's and may also include Target. (Alternative scenarios can be run, depending on different assumptions about these competitors.)

Some observers might say that Shopping Cart also competes with dollar stores and resale outlets like the Salvation Army. Assumptions about potential competition with low-end retail relies on the coarseness or fineness of the analysis that helps define the term "same industry" and "substitute goods." The analysis also lets Shopping Cart measure how much the goods are substituted, though it should be emphasized that to some degree it is also a matter of judgment. It's assumed that many previous competitors have been through a similar analysis and have

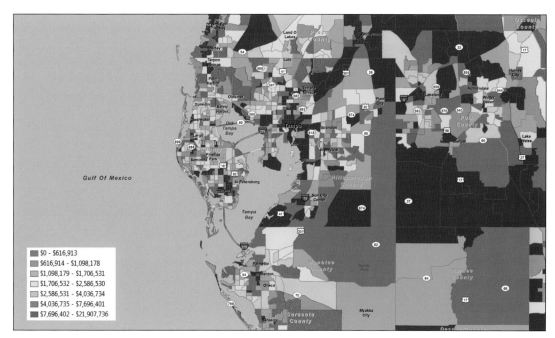

Figure 4-4. Total spending on women's apparel by census tract in the Tampa Bay Area. Data displayed in screenshots of Esri Business Analyst are courtesy of Esri; US Census Bureau; Infogroup; Bureau of Labor Statistics; Applied Geographic Solutions, Inc.; Directory of Major Malls, Inc.; GfK MRI; and Market Planning Solutions, Inc.

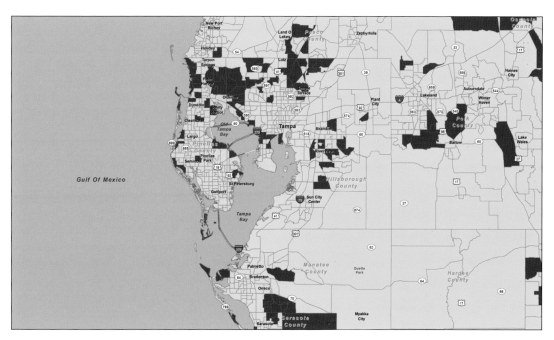

Figure 4-5. Areas (census tracts) with high spending for women's apparel (shown in green). Data displayed in screenshots of Business Analyst are courtesy of Esri; US Census Bureau; Infogroup; Bureau of Labor Statistics; Applied Geographic Solutions, Inc.; Directory of Major Malls, Inc.; GfK MRI; and Market Planning Solutions, Inc.

already located their stores strategically to maximize their trade area penetration while minimizing **cannibalization** from their own other stores.

Cannibalization is a reduction in sales in one store and the addition of sales to another store when their trade areas overlap. While the reasons stated here are not exhaustive, stores may cannibalize sales from other stores when they are more competitive with respect to prices and quality of the product or because they sell more varied goods than the cannibalized stores.

Other considerations include the costs of retail development or leased space. Viewed from the perspective of the consumer, the cost of consuming the product is the sum of the price a consumer pays plus the value of the time it takes to buy it. If the value of the time it takes to buy it is a small percentage of the total price of the good—for example, an automobile—customers might be willing to travel a great distance to make this relatively big and infrequent purchase. It also depends on the nature of the product or good in the sense that automobiles are big and need a lot of room to display them—land rents per unit are a more significant cost factor for auto retailers than for clothing boutiques.

From the perspective of the developer, lease costs mean that higher costs per square foot require higher sales per square foot. Jewelry stores tend to locate in high-cost downtown areas, while retailers selling low-cost consumer goods or groceries tend to seek lower rents consistent with the lower revenue per square foot typical for those uses. Developers must balance these considerations against decisions made by competitors in the same market.

The analysis first will look for customers who best fit the sales profile of the Shopping Cart franchise, then identify which businesses already meet the demand for goods and services that are in Shopping Cart's inventory.

BA Desktop offers sophisticated tools for site selection. It allows the analyst to identify where most customers are and distinguishes between higher probability and lower probability customer areas on the basis of income, customer profile shopping preferences, and distance—either physical or travel time—from residence to store. Because competitors are looking for the same customers, it is worthwhile to analyze first where the competitors have located.

Identifying competitors

Some firms benefit from proximity to competitors, while other firms benefit from locating at some point that is convenient to the widest customer base possible while minimizing sales losses to competitors. In the former case, for example, gas stations may thrive in close proximity to one another, engaging in price competition to attract customers. In the latter case, firms such as large shopping malls try to maximize their distance from competitors to ensure the largest share of customers for themselves.

Why do some retailers seek to cluster near one another while others seek to maximize the distance between themselves and competitors? Retail location decisions are influenced by **shopping externalities**. Shopping externalities are the sales impacts of stores on other stores in the same vicinity. Large shopping malls exist to exploit shopping externalities.

Retailers that sell essentially the same product, such as gas stations, may locate near each other primarily to capture traffic going in a particular direction, and then compete along the one dimension that can differ, the price of the product. Retailers selling complementary goods also may seek to be located near each other to take advantage of scale economies of shopping. Shoppers inclined to buy shoes may be interested in buying other clothing as well. On the other hand, retailers selling similar goods at similar prices but at greater volumes (measured by more floor space) will want to locate farther from competitors.

The map in figure 4-6 shows gas stations clustered near major intersections and along highways like US 19. The clustering at intersections occurs because stations at each corner capture enough traffic going in one direction to support the station. There is enough traffic along the highway to support multiple gas stations, and their arrangement along the highway facilitates price comparisons by travelers. While the placement of gas stations within Saint Petersburg, Florida, is strongly influenced by zoning regulations and the availability of land within mostly built-out Pinellas County, most Saint Petersburg gas stations are within a half-mile of other stations, and many are less than a quarter mile from other stations.

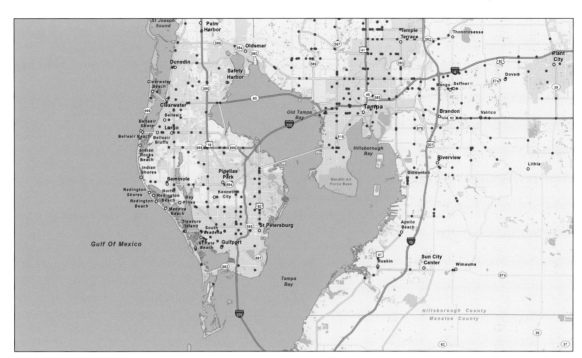

Figure 4-6. Gas stations in the Tampa Bay Area. Data displayed in screenshots of Esri Business Analyst are courtesy of Esri; US Census Bureau; Infogroup; Bureau of Labor Statistics; Applied Geographic Solutions, Inc.; Directory of Major Malls, Inc.; GfK MRI; and Market Planning Solutions, Inc.

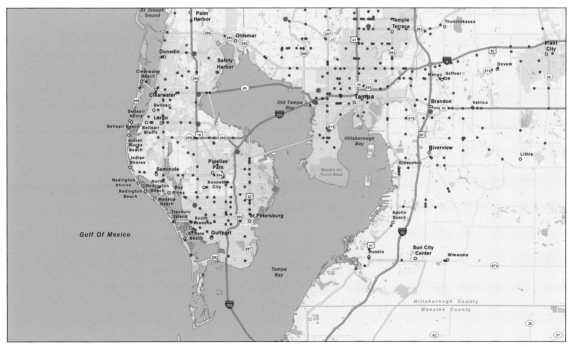

Figure 4-7. Shopping centers in the Tampa Bay Area (red dots) along with gas stations (brown dots outlined in black). Data displayed in screenshots of Esri Business Analyst are courtesy of Esri; US Census Bureau; Infogroup; Bureau of Labor Statistics; Applied Geographic Solutions, Inc.; Directory of Major Malls, Inc.; GfK MRI; and Market Planning Solutions, Inc.

Estimating trade areas

In figure 4-7, shopping malls within Pinellas County are indicated by bright red dots. Note that there are far fewer shopping malls than gas stations. Shopping malls are much farther apart than gas stations, about three miles apart on average, and dominate the retail markets for their goods within their **trade areas**.

Defining trade areas around stores

Every successful store has a catchment area from which customers are typically drawn. A trade area for a store is the catchment area where a store's attractiveness is dominant compared to other stores, and consequently, the negative effects of competitors are minimized. A store is placed to maximize the number of customers willing to travel there and not to competitors' stores. Locating a store depends on factors such as market demographics, the distance customers must travel, product differentiation, and the scale economies of the goods that are sold in stores in the same market.

For example, a store selling ski equipment in the flat and sunny state of Florida may not have many competitors in the same city because demand for winter sporting goods is going to be low. With few potential customers, there may not be any retailers specializing in these goods in the metropolitan area. On the other hand, with beautiful beaches nearby, stores specializing in marine activities and beach equipment are in high demand. While these stores could potentially be located in any commercial area within the study area, many marine-oriented stores are located near the coast because of their proximity the beach and ocean.

Stores that sell unique goods, such as luxury items, will appeal to distinct market niches such as the upper-middle and upper-class customer base. With relatively few customers, it is important to consider how many of them will purchase enough of a luxury store's stock to ensure that the store is profitable. The location criteria in this case dictate a location that is both closer to higher income areas and is more distant from other shopping malls, where comparison shopping on price may not be an advantage.

Generally, most stores draw their customers from an area around the store. These areas, called market or trade areas, can be quite small when a business such as a convenience store offers basic necessities. While also attracting customers from farther away, convenience stores attract customers who are looking for easy entry and exit with a small number of purchases. There is a balance between store size and trade area. A store with a large trade area generally needs a large and diverse set of goods to attract enough customers to survive, and vice versa. Depending upon nearby residential densities, convenience stores may have trade areas of up to a half-mile radius, while grocery stores may have trade areas with a radius of two or more miles.

Applying the Huff Model

Before choosing a store location, retailers must take into account several considerations related to the characteristics of the store, the nature of the goods it sells, and the competition it faces locally. The Huff Model formalizes the relationships among these considerations, as described in "Calibrating the Huff Model Using ArcGIS Business Analyst," a September, 2008, Esri white paper by David Huff and Brad McCallum. Basically, the Huff Model calculates the probability of shopping activity occurring at a certain store from a given geography, like a census block group or ZIP Code for which there is detailed market information. To calculate that probability (and therefore sales), the model relates measures of a store's attractiveness to shoppers as a factor that draws customers to the store against factors that repel potential customers, particularly distance from the store.

Figure 4-8 is a simple location map depicting the distribution of shopping centers in the Tampa Bay Area. Business Analyst already contains the locations of these shopping centers. The user can, with a few commands, bring up a wealth of current business data.

The earlier analysis has shown that four of these shopping centers are located in relatively wealthy areas. The four candidate shopping centers are identified in the map in figure 4-9.

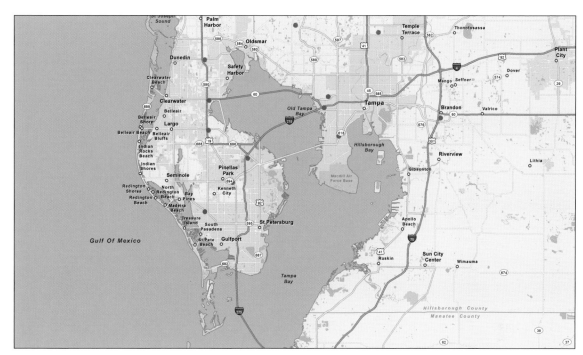

Figure 4-8. Tampa area shopping centers (red dots). Data displayed in screenshots of Esri Business Analyst are courtesy of Esri; US Census Bureau; Infogroup; Bureau of Labor Statistics; Applied Geographic Solutions, Inc.; Directory of Major Malls, Inc.; GfK MRI; and Market Planning Solutions, Inc.

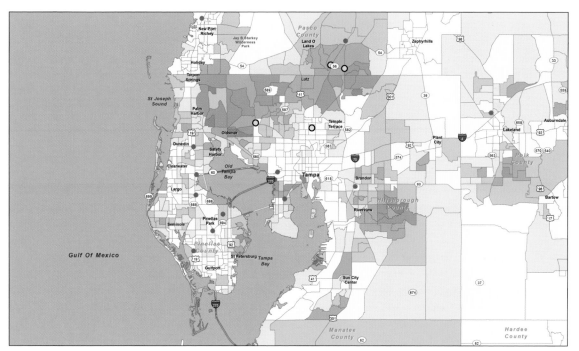

Figure 4-9. Existing shopping centers (red dots) and candidate shopping centers (yellow dots) with median income indicated by graduated colors, with the highest median income census tracts shown in darkest green. Data displayed in screenshots of Esri Business Analyst are courtesy of Esri; US Census Bureau; Infogroup; Bureau of Labor Statistics; Applied Geographic Solutions, Inc.; Directory of Major Malls, Inc.; GfK MRI; and Market Planning Solutions, Inc.

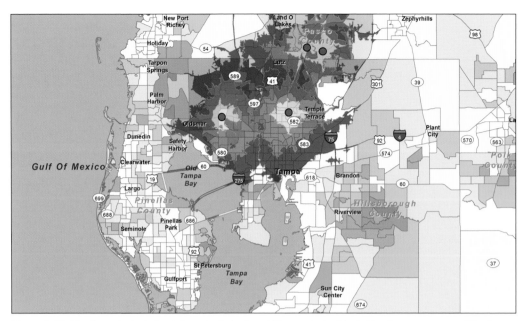

Figure 4-10. The five-minute (yellow), ten-minute (red) and fifteen-minute (blue) drive-time areas around the four candidate malls (red dots). Data displayed in screenshots of Esri Business Analyst are courtesy of Esri; US Census Bureau; Infogroup; Bureau of Labor Statistics; Applied Geographic Solutions, Inc.; Directory of Major Malls, Inc.; GfK MRI; and Market Planning Solutions, Inc.

The maps in figure 4-10 show the areas accessible for each of the candidate shopping centers at five-, ten-, and fifteen-minute drive times. Some of these drive times overlap.

Forecasting sales

The modeling tools packaged with BA Desktop are used in this section to conduct sales forecasting for the proposed store based on alternative locations. The end of this chapter includes a report generated by BA Desktop, indicating the market size included in the study area and potential sales across a variety of activities for year 2000 and forecasted for 2015.

The four candidate shopping centers identified for a possible Shopping Cart store differ by accessibility to different locations and by factors such as the size of the shopping center. These factors are proxies for the degree of shopping externalities. The original Huff Model accounts for these factors. Within BA Desktop, the Huff Model generates maps and detailed reports on sales potential at each location. (The original Huff Model can also be applied to any given location, even if there is no shopping center there, although that application is not made here.)

The attractiveness of a store to potential customers at the aggregate level (which the model must operate at) links the socioeconomic characteristics of neighborhoods to shopping profiles. As implemented in Business Analyst, the Huff Model supports analysis based not only on family income but also on many spending characteristics drawn from consumer profiles that indicate, for example, the number of cars—or even pets—in a household.

Since not all stores are created equal, the likelihood that customers will patronize a given store will vary somewhat based on the nature of the store. For example, neighborhood convenience stores have a very high decay rate for distance—meaning that their customer catchment area is relatively small compared to large shopping centers with a low decay rate and a much wider catchment area.

The results of applying the Huff Model to the four candidate shopping centers are represented in figures 4-11 to 4-14.

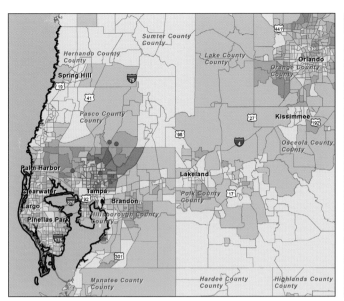

Figure 4-11. Map generated by Huff Model for mall candidate 1.

Data displayed in screenshots of Business Analyst are courtesy of Esri; US Census Bureau; Infogroup; Bureau of Labor Statistics; Applied Geographic Solutions, Inc.; Directory of Major Malls, Inc.; GfK MRI; and Market Planning Solutions, Inc.

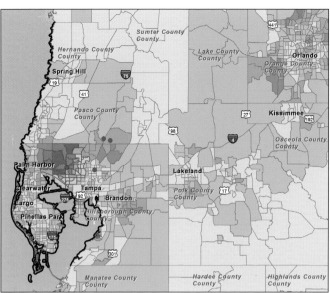

Figure 4-12. Map generated by Huff Model for mall candidate 2.

Data displayed in screenshots of Esri Business Analyst are courtesy of Esri; US Census Bureau; Infogroup; Bureau of Labor Statistics; Applied Geographic Solutions, Inc.; Directory of Major Malls, Inc.; GfK MRI; and Market Planning Solutions, Inc.

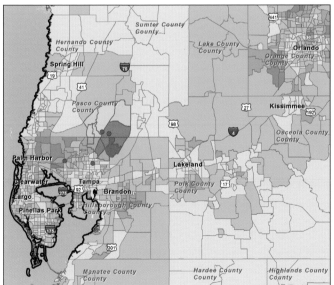

Figure 4-13. Map generated by Huff Model for mall candidate 3.

Data displayed in screenshots of Esri Business Analyst are courtesy of Esri; US Census Bureau; Infogroup; Bureau of Labor Statistics; Applied Geographic Solutions, Inc.; Directory of Major Malls, Inc.; GfK MRI; and Market Planning Solutions, Inc.

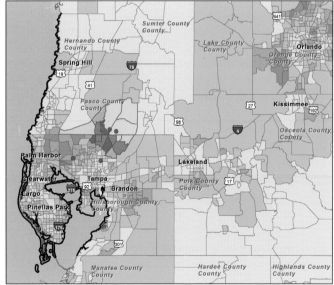

Figure 4-14. Map generated by Huff Model for mall candidate 4.

Data displayed in screenshots of Esri Business Analyst are courtesy of Esri; US Census Bureau; Infogroup; Bureau of Labor Statistics; Applied Geographic Solutions, Inc.; Directory of Major Malls, Inc.; GfK MRI; and Market Planning Solutions, Inc.

Each of these locations has different gross sales implications for Shopping Cart, as summarized in table 4-1.

Table 4-1. Gross sales of the four alternative sites for a mall

Candidate	Gross sales
Candidate 1	$1,044,434
Candidate 2	$777,506
Candidate 3	$272,704
Candidate 4	$269,727

The table tells us that the first candidate offers the greatest sales potential. However, the forecast gross sales at that location must be compared to lease rates and other cost factors before a final decision is made.

The Geographic Approach: Act

A marketing strategy for Shopping Cart

Television, radio, and the Internet are venues used by retailers to reach out to potential customers. Some retailers employ a more calibrated approach to local customers through direct mail. Using the same techniques that site selectors use to identify the optimum location for a new store, retailers target certain market segments for direct mailings of advertising materials and coupons. Direct mailing serves two related purposes—spreading awareness of a new business and overcoming potential customer inertia. This is especially the case for customers living along the seam line between two trade areas, who may be offered coupons and other select incentives to shop at the new business.

BA Desktop can be used to define trade areas by identifying the ZIP Codes for direct mailings. Each ZIP Code contained within a trade area can be blanketed with direct mailings, including ZIP Codes that fall between the boundaries of neighboring trade areas. Beyond that limit, it may not be useful to mail potential customers unless the business sending the direct mailings intends to compete on price and variety with competitors in an effort to draw short-term sales activities.

It is assumed that those activities are short-term because sales campaigns are usually limited in term. Unless customers can gain significant savings by traveling farther to shop, they prefer stores in their own trade areas over the long run. Shopping closer to home takes less time and costs less in transportation.

Other perspectives: Economic development professionals and planners

In states such as California, where local jurisdictions compete for sales tax revenue, economic development professionals may want to compete for that revenue with neighboring jurisdictions. Florida, on the other hand, provides sales tax revenue sharing among jurisdictions, and that policy limits competition among jurisdictions for shopping centers, at least on a sales tax basis. Even so, there are many other reasons to compete for shopping centers. While retail employee wages are typically low, the presence of an upscale shopping center may raise perceptions of the quality of life in the jurisdiction with the shopping center. That in turn, may have a positive effect on property values not only at the shopping center site but at neighboring properties as well. Another ripple effect is that upper-end shopping opportunities may influence the location decisions of other businesses. For example, the perception that Shopping Cart stores locate in areas with a higher quality of life is attractive to business owners and their families. It is important to remember that the issues and fiscal implications of locating a store in a particular community depend on local and state factors. As noted earlier, the implications of placing a Shopping Cart store would be different in California, for example, than in Florida.

One way to consider this question is to determine the return on investment for local government in offering tax abatements to Shopping Cart to locate within the jurisdiction. For example, a store producing $60,000 yearly in sales tax revenue and $78,000 in property taxes (at $1.30 per square foot at 60,000 square feet) would provide a return to local government of $200,000 annually, based on an additional one-fourth cent tax increase. The local jurisdiction also would also receive $160,000 in net revenue annually, based on a 20 percent tax abatement worth $40,000 dollars over ten years. From the perspective of a planner or economic development official, the potential of a Shopping Cart on local roads can be analyzed by mapping the trade area of the shopping store. Identifying where customers will come from will also identify which road segments will be most heavily traveled. (Such traffic impacts are heavier for a mall than for a single store.) The impacts suggested by Business Analyst can supplement the information provided by more formal models that analyze the traffic impacts of new projects. A secondary level of impacts from the location of a Shopping Cart store depends on where store employees decide to live. A large retailer most likely will attract employees from existing businesses in the area because the cost of relocating to the area may not be worth the increased commute time and travel costs. Still, there will be some impacts from new employment drawn to the area. Because of low prevailing wages in retail, new employees may impose costs on public facilities through increased demands for education, public safety, water, and sewer services that are passed on to all taxpayers—unless those costs are recouped through impact fees levied on the developer of the Shopping Cart store.

If low-to-moderate cost housing is scarce or concentrated in a given area, the Shopping Cart store may impose other indirect costs on a city or county. For example, transportation networks could become more crowded if employees cannot find housing close to work or if they commute from a distinct neighborhood.

Summary

Site selection is a central activity in economic development. Selecting the right site for a manufacturing facility, retail outlet, or service provider is important for the profitability of the enterprise and therefore for its long-term viability. Because these businesses are important for a community in terms of the jobs they provide, contributions to taxes, and impact on the quality of life, the public has a stake in successful site selection.

GIS generally, and BA Desktop, are tools that promote data-driven decisions about site selection. This chapter emphasized how Business Analyst, in conjunction with the Huff Model and other tools, can be used to easily make such decisions. This chapter also focused on identifying an appropriate *site* for a single store. The next chapter discusses GIS as a way to identify *areas* that will become the focus of economic development.

Appendix

This appendix presents one of the many kinds of reports that can be generated by BA Desktop. The executive summary gives an overview of the demographic and economic characteristics of the area for "Drive Time Areas Candidate 1" for the drive-time areas of five, ten, and fifteen minutes.

Executive Summary

Drive Time Areas Candidate 1
Drive Time: 5, 10, 15 minutes

Prepared By Business Analyst Desktop
Latitude: 28.19332
Longitude: -82.38816

	0 - 5 minutes	0 - 5 minutes	0 - 5 minutes
Population			
1990 Population	0	602	45,430
2000 Population	0	2,879	53,773
2010 Population	0	12,082	58,130
2015 Population	0	14,933	58,464
1990-2000 Annual Rate	0.00%	16.94%	1.70%
2000-2010 Annual Rate	0.00%	15.02%	0.76%
2010-2015 Annual Rate	0.00%	4.33%	0.11%
2010 Male Population	0.0%	48.3%	49.0%
2010 Female Population	0.0%	51.7%	51.0%
2010 Median Age	0.0	33.6	27.9

In the identified market area, the current year population is 58,130. In 2000, the Census count in the market area was 53,773. The rate of change since 2000 was 0.76 percent annually. The five-year projection for the population in the market area is 58,464, representing a change of 0.11 percent annually from 2010 to 2015. Currently, the population is 49.0 percent male and 51.0 percent female.

Population by Employment

Currently, 80.4 percent of the civilian labor force in the indentified market area is employed and 19.6 percent are unemployed. In comparison, 89.2 percent of the U.S. civilian labor force is employed, and 10.8 percent are unemployed. In five years the rate of employment in the market area will be 84.2 percent of the civilian labor force, and unemployment will be 15.8 percent. The percentage of the U.S. civilian labor force that will be employed in five years is 91.2 percent, and 8.8 percent will be unemployed. In 2000, 65.3 percent of the population aged 16 years or older in the market area participated in the labor force, and 0.1 percent were in the Armed Forces.

In the current year, the occupational distribution of the employed population is:

54.7 percent in white collar jobs (compared to 61.6 percent of the U.S. employment)
25.2 percent in service jobs (compared to 17.3 percent of U.S. employment)
20.1 percent in blue collar jobs (compared to 21.1 percent of U.S. employment)

In 2000, 68.9 percent of the market area population drove alone to work, and 1.1 percent worked at home. The average travel time to work in 2000 was 26.8 minutes in the market area, compared to the U.S average of 25.5 minutes.

Population by Education

In the current year, the educational attainment of the population aged 25 years or older in the market area was distributed as follows:
20.1 percent had not earned a high school diploma (14.8 percent in the U.S)
33.4 percent were high school graduates only (29.6 percent in the U.S.)
9.4 percent had completed an Associate degree (7.7 percent in the U.S.)
12.4 percent had a Bachelor's degree (17.7 percent in the U.S.)
6.6 percent had earned a Master's/Professional/Doctorate Degree (10.4 percent in the U.S.)

Per Capita Income

	0 - 5 minutes	0 - 5 minutes	0 - 5 minutes
1990 Per Capita Income	$0	$15,118	$9,602
2000 Per Capita Income	$0	$28,945	$13,129
2010 Per Capita Income	$0	$31,922	$15,967
2015 Per Capita Income	$0	$33,412	$17,977
1990-2000 Annual Rate	0.00%	6.71%	3.18%
2000-2010 Annual Rate	0.00%	0.96%	1.93%
2010-2015 Annual Rate	0.00%	0.92%	2.40%
Households			
1990 Households	0	232	18,546
2000 Households	0	1,037	21,287
2010 Total Households	0	4,108	22,685
2015 Total Households	0	5,087	22,835
1990-2000 Annual Rate	0.00%	16.15%	1.39%
2000-2010 Annual Rate	0.00%	14.37%	0.62%
2010-2015 Annual Rate	0.00%	4.37%	0.13%
2010 Average Household Size	0.00	2.94	2.35

The household count in this market area has changed from 21,287 in 2000 to 22,685 in the current year, a change of 0.62 percent annually. The five-year projection of households is 22,835, a change of 0.13 percent annually from the current year total. Average household size is currently 2.35, compared to 2.32 in the year 2000. The number of families in the current year is 11,083 in the market area.

Data Note: Income is expressed in current dollars
Source: U.S. Bureau and Census, 2000 Census of Population and Housing, ESRI forecast for 2010 and 2015. ESRI converted 1990 Census data into 2000 geography.

August 09, 2011

Made with ESRI Business Analyst

©2010 ESRI   www.esri.com/ba   800-447-9778 Try it Now!   Page 1 of 10

 **esri**

Drive Time Areas Candidate 1
Drive Time: 5, 10, 15 minutes

Prepared By Business Analyst Desktop
Latitude: 28.0583
Longitude: -82.4346

	0 - 5 minutes	0 - 5 minutes	0 - 5 minutes
Households by Income			

Current median household income is $29,313 in the market area, compared to $54,442 for all U.S. households. Median household income is projected to be $32,549 in five years. In 2000, median household income was $24,276.

Current average household income is $37,212 in this market area, compared to $70,173 for all U.S households. Average household income is projected to be $41,828 in five years. In 2000, average household income was $31,330, compared to $21,941 in 1990.

Current per capita income is $15,967 in the market area, compared to the U.S. per capita income of $26,739. The per capita income is projected to be $17,977 in five years. In 2000, the per capita income was $13,129, compared to $9,602 in 1990.

	0 - 5 minutes	0 - 5 minutes	0 - 5 minutes
Median Household Income			
2000 Median Household Income	$0	$64,039	$24,276
2010 Median Household Income	$0	$75,835	$29,313
2015 Median Household Income	$0	$77,341	$32,549
2000-2010 Annual Rate	0.00%	1.66%	1.86%
2010-2015 Annual Rate	0.00%	0.39%	2.12%
Average Household Income			
1990 Average Household Income	$0	$36,623	$21,941
2000 Average Household Income	$0	$78,475	$31,330
2010 Average Household Income	$0	$88,535	$37,212
2015 Average Household Income	$0	$92,476	$41,828
1990-2000 Annual Rate	0.00%	7.92%	3.63%
2000-2010 Annual Rate	0.00%	1.18%	1.69%
2010-2015 Annual Rate	0.00%	0.87%	2.37%
2010 Housing			
1990 Total Housing Units	0	303	23,768
2000 Total Housing Units	0	1,068	23,899
2010 Total Housing Units	0	4,595	27,044
2015 Total Housing Units	0	5,833	28,033
1990 Owner Occupied Housing Units	0	207	5,549
1990 Renter Occupied Housing Units	0	24	12,998
1990 Vacant Housing Units	0	86	5,178
2000 Owner Occupied Housing Units	0	997	5,763
2000 Renter Occupied Housing Units	0	40	15,524
2000 Vacant Housing Units	0	97	2,571
2010 Owner Occupied Housing Units	0	3,694	6,134
2010 Renter Occupied Housing Units	0	414	16,551
2010 Vacant Housing Units	0	487	4,359
2015 Owner Occupied Housing Units	0	4,554	6,143
2015 Renter Occupied Housing Units	0	534	16,693
2015 Vacant Housing Units	0	746	5,198

Currently, 22.7 percent of the 27,044 housing units in the market area are owner occupied; 61.2 percent, renter occupied; and 16.1 are vacant. In 2000, there were 23,899 housing units - 24.1 percent owner occupied, 65.0. percent renter occupied, and 10.8 percent vacant. The rate of change in housing units since 2000 is 1.21 percent. Median home value in the market area is $84,388, compared to a median home value of $157,913 for the U.S. In five years, median value is projected to change by 2.82 percent annually to $96,976. From 2000 to the current year, median home value change by 2.76 percent annually.

Data Note: Income is expressed in current dollars
Source: U.S. Bureau and Census, 2000 Census of Population and Housing, ESRI forecast for 2010 and 2015. ESRI converted 1990 Census data into 2000 geography.

August 09, 2011

Made with ESRI Business Analyst

©2010 ESRI

www.esri.com/ba   800-447-9778   Try it Now!

Page 2 of 10

 **esri**

Executive Summary

Drive Time Areas Candidate 1
Drive Time: 5, 10, 15 minutes

Prepared By Business Analyst Desktop
Latitude: 28.0691
Longitude: -82.5768

	0 - 5 minutes	0 - 10 minutes	0 - 10 minutes
Population			
1990 Population	6,321	549	6,659
2000 Population	10,586	2,386	18,818
2010 Population	15,475	6,785	39,200
2015 Population	16,562	8,003	46,553
1990-2000 Annual Rate	5.29%	15.83%	10.95%
2000-2010 Annual Rate	3.77%	10.73%	7.42%
2010-2015 Annual Rate	1.37%	3.36%	3.50%
2010 Male Population	48.7%	49.9%	48.8%
2010 Female Population	51.3%	50.1%	51.2%
2010 Median Age	35.5	33.5	34.0

In the identified market area, the current year population is 39,200. In 2000, the Census count in the market area was 18,818. The rate of change since 2000 was 7.42 percent annually. The five-year projection for the population in the market area is 46,553, representing a change of 3.50 percent annually from 2010 to 2015. Currently, the population is 48.8 percent male and 51.2 percent female.

Population by Employment

Currently, 91.5 percent of the civilian labor force in the indentified market area is employed and 8.5 percent are unemployed. In comparison, 89.2 percent of the U.S. civilian labor force is employed, and 10.8 percent are unemployed. In five years the rate of employment in the market area will be 93.4 percent of the civilian labor force, and unemployment will be 6.6 percent. The percentage of the U.S. civilian labor force that will be employed in five years is 91.2 percent, and 8.8 percent will be unemployed. In 2000, 72.4 percent of the population aged 16 years or older in the market area participated in the labor force, and 0.2 percent were in the Armed Forces.

In the current year, the occupational distribution of the employed population is:

68.7 percent in white collar jobs (compared to 61.6 percent of the U.S. employment)
15.9 percent in service jobs (compared to 17.3 percent of U.S. employment)
15.4 percent in blue collar jobs (compared to 21.1 percent of U.S. employment)

In 2000, 84.9 percent of the market area population drove alone to work, and 3.2 percent worked at home. The average travel time to work in 2000 was 32.5 minutes in the market area, compared to the U.S average of 25.5 minutes.

Population by Education

In the current year, the educational attainment of the population aged 25 years or older in the market area was distributed as follows:

4.2 percent had not earned a high school diploma (14.8 percent in the U.S)
18.7 percent were high school graduates only (29.6 percent in the U.S.)
11.0 percent had completed an Associate degree (7.7 percent in the U.S.)
32.1 percent had a Bachelor's degree (17.7 percent in the U.S.)
14.4 percent had earned a Master's/Professional/Doctorate Degree (10.4 percent in the U.S.)

Per Capita Income			
1990 Per Capita Income	$14,416	$25,325	$17,210
2000 Per Capita Income	$23,857	$31,618	$26,803
2010 Per Capita Income	$31,286	$41,407	$32,790
2015 Per Capita Income	$33,456	$44,608	$35,350
1990-2000 Annual Rate	5.17%	2.24%	4.53%
2000-2010 Annual Rate	2.68%	2.67%	1.99%
2010-2015 Annual Rate	1.35%	1.50%	1.51%
Households			
1990 Households	2,202	241	2,426
2000 Households	3,895	1,071	6,804
2010 Total Households	5,712	3,103	13,764
2015 Total Households	6,149	3,672	16,334
1990-2000 Annual Rate	5.87%	16.09%	10.86%
2000-2010 Annual Rate	3.81%	10.94%	7.12%
2010-2015 Annual Rate	1.49%	3.42%	3.48%
2010 Average Household Size	2.70	2.19	2.85

The household count in this market area has changed from 6,804 in 2000 to 13,764 in the current year, a change of 7.12 percent annually. The five-year projection of households is 16,334, a change of 3.48 percent annually from the current year total. Average household size is currently 2.85, compared to 2.77 in the year 2000. The number of families in the current year is 10,286 in the market area.

Data Note: Income is expressed in current dollars
Source: U.S. Bureau and Census, 2000 Census of Population and Housing, ESRI forecast for 2010 and 2015. ESRI converted 1990 Census data into 2000 geography.

August 09, 2011

Drive Time Areas Candidate 1
Drive Time: 5, 10, 15 minutes

Prepared By Business Analyst Desktop
Latitude: 28.186
Longitude: -82.3539

	0 - 5 minutes	0 - 10 minutes	0 - 10 minutes
Households by Income			

Current median household income is $77,593 in the market area, compared to $54,442 for all U.S. households. Median household income is projected to be $81,092 in five years. In 2000, median household income was $62,165.

Current average household income is $90,410 in this market area, compared to $70,173 for all U.S households. Average household income is projected to be $97,507 in five years. In 2000, average household income was $72,934, compared to $46,621 in 1990.

Current per capita income is $32,790 in the market area, compared to the U.S. per capita income of $26,739. The per capita income is projected to be $35,350 in five years. In 2000, the per capita income was $26,803, compared to $17,210 in 1990.

	0 - 5 minutes	0 - 10 minutes	0 - 10 minutes
Median Household Income			
2000 Median Household Income	$56,162	$61,188	$62,165
2010 Median Household Income	$72,346	$86,981	$77,593
2015 Median Household Income	$75,981	$98,272	$81,092
2000-2010 Annual Rate	2.50%	3.49%	2.19%
2010-2015 Annual Rate	0.99%	2.47%	0.89%
Average Household Income			
1990 Average Household Income	$42,114	$59,192	$46,621
2000 Average Household Income	$66,577	$74,859	$72,934
2010 Average Household Income	$86,852	$99,477	$90,410
2015 Average Household Income	$92,411	$106,390	$97,507
1990-2000 Annual Rate	4.69%	2.38%	4.58%
2000-2010 Annual Rate	2.63%	2.81%	2.12%
2010-2015 Annual Rate	1.25%	1.35%	1.52%
2010 Housing			
1990 Total Housing Units	2,402	334	2,787
2000 Total Housing Units	4,315	1,594	7,190
2010 Total Housing Units	6,637	4,464	15,332
2015 Total Housing Units	7,239	5,387	18,518
1990 Owner Occupied Housing Units	1,834	131	2,183
1990 Renter Occupied Housing Units	368	110	243
1990 Vacant Housing Units	184	94	397
2000 Owner Occupied Housing Units	3,102	518	5,737
2000 Renter Occupied Housing Units	793	553	1,067
2000 Vacant Housing Units	441	441	627
2010 Owner Occupied Housing Units	4,311	1,837	11,454
2010 Renter Occupied Housing Units	1,400	1,265	2,310
2010 Vacant Housing Units	926	1,361	1,568
2015 Owner Occupied Housing Units	4,591	2,173	13,638
2015 Renter Occupied Housing Units	1,558	1,499	2,697
2015 Vacant Housing Units	1,090	1,715	2,183

Currently, 74.7 percent of the 15,332 housing units in the market area are owner occupied; 15.1 percent, renter occupied; and 10.2 are vacant. In 2000, there were 7,190 housing units - 79.8 percent owner occupied, 14.8. percent renter occupied, and 8.7 percent vacant. The rate of change in housing units since 2000 is 7.67 percent. Median home value in the market area is $171,654, compared to a median home value of $157,913 for the U.S. In five years, median value is projected to change by 3.16 percent annually to $200,515. From 2000 to the current year, median home value change by 2.90 percent annually.

Data Note: Income is expressed in current dollars
Source: U.S. Bureau and Census, 2000 Census of Population and Housing, ESRI forecast for 2010 and 2015. ESRI converted 1990 Census data into 2000 geography.

August 09, 2011

Made with ESRI Business Analyst

©2010 ESRI

www.esri.com/ba   800-447-9778   Try it Now!

Part II: Applying the Geographic Approach and GIS to economic development analysis

 **esri**

Executive Summary

Drive Time Areas Candidate 1
Drive Time: 5, 10, 15 minutes

Prepared By Business Analyst Desktop
Latitude: 28.0691
Longitude: -82.5768

	0 - 10 minutes	0 - 10 minutes	0 - 15 minutes
Population			
1990 Population	95,053	186,680	322,315
2000 Population	122,950	207,436	346,470
2010 Population	141,326	218,053	367,707
2015 Population	144,828	217,493	368,092
1990-2000 Annual Rate	2.61%	1.06%	0.73%
2000-2010 Annual Rate	1.37%	0.49%	0.58%
2010-2015 Annual Rate	0.49%	-0.05%	0.02%
2010 Male Population	48.4%	48.7%	48.6%
2010 Female Population	51.6%	51.3%	51.4%
2010 Median Age	36.7	32.9	34.3

In the identified market area, the current year population is 367,707. In 2000, the Census count in the market area was 346,470. The rate of change since 2000 was 0.58 percent annually. The five-year projection for the population in the market area is 368,092, representing a change of 0.02 percent annually from 2010 to 2015. Currently, the population is 48.6 percent male and 51.4 percent female.

Population by Employment

Currently, 84.2 percent of the civilian labor force in the indentified market area is employed and 15.8 percent are unemployed. In comparison, 89.2 percent of the U.S. civilian labor force is employed, and 10.8 percent are unemployed. In five years the rate of employment in the market area will be 87.4 percent of the civilian labor force, and unemployment will be 12.6 percent. The percentage of the U.S. civilian labor force that will be employed in five years is 91.2 percent, and 8.8 percent will be unemployed. In 2000, 64.6 percent of the population aged 16 years or older in the market area participated in the labor force, and 0.2 percent were in the Armed Forces.

In the current year, the occupational distribution of the employed population is:

  62.9 percent in white collar jobs (compared to 61.6 percent of the U.S. employment)
  19.5 percent in service jobs (compared to 17.3 percent of U.S. employment)
  17.6 percent in blue collar jobs (compared to 21.1 percent of U.S. employment)

In 2000, 76.4 percent of the market area population drove alone to work, and 2.3 percent worked at home. The average travel time to work in 2000 was 24.2 minutes in the market area, compared to the U.S average of 25.5 minutes.

Population by Education

In the current year, the educational attainment of the population aged 25 years or older in the market area was distributed as follows:

  17.8 percent had not earned a high school diploma (14.8 percent in the U.S)
  28.8 percent were high school graduates only (29.6 percent in the U.S.)
  9.0 percent had completed an Associate degree (7.7 percent in the U.S.)
  17.4 percent had a Bachelor's degree (17.7 percent in the U.S.)
  9.7 percent had earned a Master's/Professional/Doctorate Degree (10.4 percent in the U.S.)

Per Capita Income			
1990 Per Capita Income	$15,856	$12,317	$12,535
2000 Per Capita Income	$23,314	$17,685	$19,008
2010 Per Capita Income	$29,427	$21,240	$22,724
2015 Per Capita Income	$32,189	$23,685	$25,395
1990-2000 Annual Rate	3.93%	3.68%	4.25%
2000-2010 Annual Rate	2.30%	1.80%	1.76%
2010-2015 Annual Rate	1.81%	2.20%	2.25%
Households			
1990 Households	35,060	76,328	131,917
2000 Households	46,295	84,230	142,064
2010 Total Households	53,330	88,609	151,482
2015 Total Households	54,796	88,701	152,310
1990-2000 Annual Rate	2.82%	0.99%	0.74%
2000-2010 Annual Rate	1.39%	0.50%	0.63%
2010-2015 Annual Rate	0.54%	0.02%	0.11%
2010 Average Household Size	2.64	2.38	2.36

The household count in this market area has changed from 142,064 in 2000 to 151,482 in the current year, a change of 0.63 percent annually. The five-year projection of households is 152,310, a change of 0.11 percent annually from the current year total. Average household size is currently 2.36, compared to 2.37 in the year 2000. The number of families in the current year is 82,192 in the market area.

Data Note: Income is expressed in current dollars
Source: U.S. Bureau and Census, 2000 Census of Population and Housing, ESRI forecast for 2010 and 2015. ESRI converted 1990 Census data into 2000 geography.

August 09, 2011

Made with ESRI Business Analyst

Executive Summary

Drive Time Areas Candidate 1
Drive Time: 5, 10, 15 minutes

Prepared By Business Analyst Desktop
Latitude: 28.0583
Longitude: -82.4346

	0 - 10 minutes	0 - 10 minutes	0 - 15 minutes
Households by Income			

Current median household income is $40,179 in the market area, compared to $54,442 for all U.S. households. Median household income is projected to be $47,714 in five years. In 2000, median household income was $31,695.

Current average household income is $54,115 in this market area, compared to $70,173 for all U.S households. Average household income is projected to be $60,183 in five years. In 2000, average household income was $45,613, compared to $30,202 in 1990.

Current per capita income is $22,724 in the market area, compared to the U.S. per capita income of $26,739. The per capita income is projected to be $25,395 in five years. In 2000, the per capita income was $19,008, compared to $12,535 in 1990.

	0 - 10 minutes	0 - 10 minutes	0 - 15 minutes
Median Household Income			
2000 Median Household Income	$49,872	$31,015	$31,695
2010 Median Household Income	$63,628	$38,906	$40,179
2015 Median Household Income	$67,902	$45,890	$47,714
2000-2010 Annual Rate	2.40%	2.24%	2.34%
2010-2015 Annual Rate	1.31%	3.36%	3.50%
Average Household Income			
1990 Average Household Income	$42,607	$29,750	$30,202
2000 Average Household Income	$61,861	$42,870	$45,613
2010 Average Household Income	$77,981	$51,183	$54,115
2015 Average Household Income	$85,084	$56,827	$60,183
1990-2000 Annual Rate	3.80%	3.72%	4.21%
2000-2010 Annual Rate	2.28%	1.74%	1.68%
2010-2015 Annual Rate	1.76%	2.11%	2.15%
2010 Housing			
1990 Total Housing Units	38,431	89,544	152,186
2000 Total Housing Units	49,094	91,779	155,076
2010 Total Housing Units	58,492	101,231	173,480
2015 Total Housing Units	61,052	103,884	178,752
1990 Owner Occupied Housing Units	24,587	38,779	67,270
1990 Renter Occupied Housing Units	10,473	37,550	64,647
1990 Vacant Housing Units	3,389	13,182	20,163
2000 Owner Occupied Housing Units	34,078	41,396	71,246
2000 Renter Occupied Housing Units	12,216	42,834	70,819
2000 Vacant Housing Units	2,816	7,527	12,900
2010 Owner Occupied Housing Units	38,005	42,052	73,672
2010 Renter Occupied Housing Units	15,324	46,558	77,810
2010 Vacant Housing Units	5,163	12,621	21,999
2015 Owner Occupied Housing Units	38,904	41,658	73,493
2015 Renter Occupied Housing Units	15,891	47,043	78,817
2015 Vacant Housing Units	6,256	15,183	26,442

Currently, 42.5 percent of the 173,480 housing units in the market area are owner occupied; 44.9 percent, renter occupied; and 12.7 are vacant. In 2000, there were 155,076 housing units - 45.9 percent owner occupied, 45.7. percent renter occupied, and 8.3 percent vacant. The rate of change in housing units since 2000 is 1.10 percent. Median home value in the market area is $102,990, compared to a median home value of $157,913 for the U.S. In five years, median value is projected to change by 3.00 percent annually to $119,388. From 2000 to the current year, median home value change by 2.74 percent annually.

Data Note: Income is expressed in current dollars
Source: U.S. Bureau and Census, 2000 Census of Population and Housing, ESRI forecast for 2010 and 2015. ESRI converted 1990 Census data into 2000 geography.

August 09, 2011

Made with ESRI Business Analyst

©2010 ESRI   www.esri.com/ba   800-447-9778   Try it Now!

Part II: Applying the Geographic Approach and GIS to economic development analysis

Executive Summary

Drive Time Areas Candidate 1
Drive Time: 5, 10, 15 minutes

Prepared By Business Analyst Desktop
Latitude: 28.19332
Longitude: -82.38816

	0 - 15 minutes	0 - 15 minutes	0 - 15 minutes
Population			
1990 Population	1,948	15,429	167,147
2000 Population	4,619	41,488	210,522
2010 Population	12,304	85,462	243,508
2015 Population	14,461	100,704	250,535
1990-2000 Annual Rate	9.02%	10.40%	2.33%
2000-2010 Annual Rate	10.03%	7.30%	1.43%
2010-2015 Annual Rate	3.28%	3.34%	0.57%
2010 Male Population	49.9%	48.8%	48.7%
2010 Female Population	50.1%	51.2%	51.3%
2010 Median Age	34.1	34.9	37.6

In the identified market area, the current year population is 243,508. In 2000, the Census count in the market area was 210,522. The rate of change since 2000 was 1.43 percent annually. The five-year projection for the population in the market area is 250,535, representing a change of 0.57 percent annually from 2010 to 2015. Currently, the population is 48.7 percent male and 51.3 percent female.

Population by Employment

Currently, 88.8 percent of the civilian labor force in the indentified market area is employed and 11.2 percent are unemployed. In comparison, 89.2 percent of the U.S. civilian labor force is employed, and 10.8 percent are unemployed. In five years the rate of employment in the market area will be 91.2 percent of the civilian labor force, and unemployment will be 8.8 percent. The percentage of the U.S. civilian labor force that will be employed in five years is 91.2 percent, and 8.8 percent will be unemployed. In 2000, 71.1 percent of the population aged 16 years or older in the market area participated in the labor force, and 0.4 percent were in the Armed Forces.

In the current year, the occupational distribution of the employed population is:

72.3 percent in white collar jobs (compared to 61.6 percent of the U.S. employment)

14.1 percent in service jobs (compared to 17.3 percent of U.S. employment)

13.6 percent in blue collar jobs (compared to 21.1 percent of U.S. employment)

In 2000, 81.7 percent of the market area population drove alone to work, and 3.4 percent worked at home. The average travel time to work in 2000 was 26.9 minutes in the market area, compared to the U.S average of 25.5 minutes.

Population by Education

In the current year, the educational attainment of the population aged 25 years or older in the market area was distributed as follows:

9.6 percent had not earned a high school diploma (14.8 percent in the U.S)

23.5 percent were high school graduates only (29.6 percent in the U.S.)

11.2 percent had completed an Associate degree (7.7 percent in the U.S.)

24.9 percent had a Bachelor's degree (17.7 percent in the U.S.)

11.4 percent had earned a Master's/Professional/Doctorate Degree (10.4 percent in the U.S.)

Per Capita Income			
1990 Per Capita Income	$20,805	$15,562	$17,464
2000 Per Capita Income	$31,238	$27,066	$25,463
2010 Per Capita Income	$43,137	$32,372	$31,118
2015 Per Capita Income	$46,734	$34,441	$34,454
1990-2000 Annual Rate	4.15%	5.69%	3.84%
2000-2010 Annual Rate	3.20%	1.76%	1.98%
2010-2015 Annual Rate	1.61%	1.25%	2.06%
Households			
1990 Households	727	5,587	64,419
2000 Households	1,830	15,097	82,199
2010 Total Households	4,925	30,782	95,099
2015 Total Households	5,806	36,334	98,062
1990-2000 Annual Rate	9.67%	10.45%	2.47%
2000-2010 Annual Rate	10.14%	7.20%	1.43%
2010-2015 Annual Rate	3.35%	3.37%	0.62%
2010 Average Household Size	2.50	2.78	2.55

The household count in this market area has changed from 82,199 in 2000 to 95,099 in the current year, a change of 1.43 percent annually. The five-year projection of households is 98,062, a change of 0.62 percent annually from the current year total. Average household size is currently 2.55, compared to 2.55 in the year 2000. The number of families in the current year is 63,684 in the market area.

Data Note: Income is expressed in current dollars
Source: U.S. Bureau and Census, 2000 Census of Population and Housing, ESRI forecast for 2010 and 2015. ESRI converted 1990 Census data into 2000 geography.

August 09, 2011

Made with ESRI Business Analyst

Executive Summary

Drive Time Areas Candidate 1
Drive Time: 5, 10, 15 minutes

Prepared By Business Analyst Desktop
Latitude: 28.0691
Longitude: -82.5768

	0 - 15 minutes	0 - 15 minutes	0 - 15 minutes
Households by Income			

Current median household income is $62,737 in the market area, compared to $54,442 for all U.S. households. Median household income is projected to be $67,611 in five years. In 2000, median household income was $48,836.

Current average household income is $79,480 in this market area, compared to $70,173 for all U.S households. Average household income is projected to be $87,804 in five years. In 2000, average household income was $64,924, compared to $45,364 in 1990.

Current per capita income is $31,118 in the market area, compared to the U.S. per capita income of $26,739. The per capita income is projected to be $34,454 in five years. In 2000, the per capita income was $25,463, compared to $17,464 in 1990.

Median Household Income			
2000 Median Household Income	$62,022	$61,980	$48,836
2010 Median Household Income	$90,524	$75,741	$62,737
2015 Median Household Income	$100,951	$78,437	$67,611
2000-2010 Annual Rate	3.76%	1.98%	2.47%
2010-2015 Annual Rate	2.20%	0.70%	1.51%
Average Household Income			
1990 Average Household Income	$50,000	$42,500	$45,364
2000 Average Household Income	$78,962	$73,714	$64,924
2010 Average Household Income	$112,323	$89,464	$79,480
2015 Average Household Income	$121,268	$95,077	$87,804
1990-2000 Annual Rate	4.68%	5.66%	3.65%
2000-2010 Annual Rate	3.50%	1.91%	1.99%
2010-2015 Annual Rate	1.54%	1.22%	2.01%
2010 Housing			
1990 Total Housing Units	909	6,342	70,886
2000 Total Housing Units	2,409	16,529	87,399
2010 Total Housing Units	6,498	34,887	104,858
2015 Total Housing Units	7,790	41,961	109,878
1990 Owner Occupied Housing Units	528	4,800	42,593
1990 Renter Occupied Housing Units	199	786	21,826
1990 Vacant Housing Units	181	766	6,467
2000 Owner Occupied Housing Units	1,116	12,954	56,776
2000 Renter Occupied Housing Units	714	2,143	25,423
2000 Vacant Housing Units	500	1,454	5,200
2010 Owner Occupied Housing Units	3,328	25,916	64,385
2010 Renter Occupied Housing Units	1,597	4,866	30,714
2010 Vacant Housing Units	1,573	4,105	9,759
2015 Owner Occupied Housing Units	3,929	30,555	66,575
2015 Renter Occupied Housing Units	1,877	5,779	31,487
2015 Vacant Housing Units	1,984	5,627	11,816

Currently, 61.4 percent of the 104,858 housing units in the market area are owner occupied; 29.3 percent, renter occupied; and 9.3 are vacant. In 2000, there were 87,399 housing units - 65.0 percent owner occupied, 29.1 percent renter occupied, and 5.9 percent vacant. The rate of change in housing units since 2000 is 1.79 percent. Median home value in the market area is $145,314, compared to a median home value of $157,913 for the U.S. In five years, median value is projected to change by 2.92 percent annually to $167,842. From 2000 to the current year, median home value change by 3.28 percent annually.

Data Note: Income is expressed in current dollars
Source: U.S. Bureau and Census, 2000 Census of Population and Housing, ESRI forecast for 2010 and 2015. ESRI converted 1990 Census data into 2000 geography.

August 09, 2011

Made with ESRI Business Analyst

©2010 ESRI

www.esri.com/ba   800-447-9778 Try it Now!

Page 8 of 10

Executive Summary

	Whole Layer (Drive Time Areas
Population	
1990 Population	848,134
2000 Population	1,021,926
2010 Population	1,200,033
2015 Population	1,240,629
1990-2000 Annual Rate	1.88%
2000-2010 Annual Rate	1.58%
2010-2015 Annual Rate	0.67%
2010 Male Population	48.7%
2010 Female Population	51.3%
2010 Median Age	34.7

In the identified market area, the current year population is 1,200,033. In 2000, the Census count in the market area was 1,021,926. The rate of change since 2000 was 1.58 percent annually. The five-year projection for the population in the market area is 1,240,629, representing a change of 0.67 percent annually from 2010 to 2015. Currently, the population is 48.7 percent male and 51.3 percent female.

Population by Employment

Currently, 86.5 percent of the civilian labor force in the indentified market area is employed and 13.5 percent are unemployed. In comparison, 89.2 percent of the U.S. civilian labor force is employed, and 10.8 percent are unemployed. In five years the rate of employment in the market area will be 89.4 percent of the civilian labor force, and unemployment will be 10.6 percent. The percentage of the U.S. civilian labor force that will be employed in five years is 91.2 percent, and 8.8 percent will be unemployed. In 2000, 67.7 percent of the population aged 16 years or older in the market area participated in the labor force, and 0.2 percent were in the Armed Forces.

In the current year, the occupational distribution of the employed population is:

66.6 percent in white collar jobs (compared to 61.6 percent of the U.S. employment)

17.4 percent in service jobs (compared to 17.3 percent of U.S. employment)

16.1 percent in blue collar jobs (compared to 21.1 percent of U.S. employment)

In 2000, 78.7 percent of the market area population drove alone to work, and 2.6 percent worked at home. The average travel time to work in 2000 was 26.2 minutes in the market area, compared to the U.S average of 25.5 minutes.

Population by Education

In the current year, the educational attainment of the population aged 25 years or older in the market area was distributed as follows:

13.2 percent had not earned a high school diploma (14.8 percent in the U.S)

26.2 percent were high school graduates only (29.6 percent in the U.S.)

10.3 percent had completed an Associate degree (7.7 percent in the U.S.)

21.3 percent had a Bachelor's degree (17.7 percent in the U.S.)

10.5 percent had earned a Master's/Professional/Doctorate Degree (10.4 percent in the U.S.)

Per Capita Income	
1990 Per Capita Income	$13,810
2000 Per Capita Income	$20,915
2010 Per Capita Income	$26,154
2015 Per Capita Income	$29,053
1990-2000 Annual Rate	4.24%
2000-2010 Annual Rate	2.20%
2010-2015 Annual Rate	2.12%
Households	
1990 Households	337,685
2000 Households	405,808
2010 Total Households	473,598
2015 Total Households	490,087
1990-2000 Annual Rate	1.85%
2000-2010 Annual Rate	1.52%
2010-2015 Annual Rate	0.69%
2010 Average Household Size	2.48

The household count in this market area has changed from 405,808 in 2000 to 473,598 in the current year, a change of 1.52 percent annually. The five-year projection of households is 490,087, a change of 0.69 percent annually from the current year total. Average household size is currently 2.48, compared to 2.47 in the year 2000. The number of families in the current year is 288,415 in the market area.

Data Note: Income is expressed in current dollars
Source: U.S. Bureau and Census, 2000 Census of Population and Housing, ESRI forecast for 2010 and 2015. ESRI converted 1990 Census data into 2000 geography.

August 09, 2011

Made with ESRI Business Analyst

©2010 ESRI   www.esri.com/ba   800-447-9778 Try it Now!   Page 9 of 10

Whole Layer (Drive Time Areas

Households by Income

Current median household income is $50,473 in the market area, compared to $54,442 for all U.S. households. Median household income is projected to be $57,731 in five years. In 2000, median household income was $37,797.

Current average household income is $65,485 in this market area, compared to $70,173 for all U.S households. Average household income is projected to be $72,672 in five years. In 2000, average household income was $52,085, compared to $34,293 in 1990.

Current per capita income is $26,154 in the market area, compared to the U.S. per capita income of $26,739. The per capita income is projected to be $29,053 in five years. In 2000, the per capita income was $20,915, compared to $13,810 in 1990.

Median Household Income

2000 Median Household Income	$37,797
2010 Median Household Income	$50,473
2015 Median Household Income	$57,731
2000-2010 Annual Rate	2.86%
2010-2015 Annual Rate	2.72%

Average Household Income

1990 Average Household Income	$34,293
2000 Average Household Income	$52,085
2010 Average Household Income	$65,485
2015 Average Household Income	$72,672
1990-2000 Annual Rate	4.27%
2000-2010 Annual Rate	2.26%
2010-2015 Annual Rate	2.10%

2010 Housing

1990 Total Housing Units	387,890
2000 Total Housing Units	440,351
2010 Total Housing Units	537,519
2015 Total Housing Units	568,327
1990 Owner Occupied Housing Units	188,460
1990 Renter Occupied Housing Units	149,225
1990 Vacant Housing Units	50,087
2000 Owner Occupied Housing Units	233,683
2000 Renter Occupied Housing Units	172,126
2000 Vacant Housing Units	34,573
2010 Owner Occupied Housing Units	274,788
2010 Renter Occupied Housing Units	198,811
2010 Vacant Housing Units	63,920
2015 Owner Occupied Housing Units	286,213
2015 Renter Occupied Housing Units	203,874
2015 Vacant Housing Units	78,240

Currently, 51.1 percent of the 537,519 housing units in the market area are owner occupied; 37.0 percent, renter occupied; and 11.9 are vacant. In 2000, there were 440,351 housing units - 53.1 percent owner occupied, 39.1. percent renter occupied, and 7.9 percent vacant. The rate of change in housing units since 2000 is 1.96 percent. Median home value in the market area is $125,303, compared to a median home value of $157,913 for the U.S. In five years, median value is projected to change by 3.27 percent annually to $147,197. From 2000 to the current year, median home value change by 3.25 percent annually.

Data Note: Income is expressed in current dollars
Source: U.S. Bureau and Census, 2000 Census of Population and Housing, ESRI forecast for 2010 and 2015. ESRI converted 1990 Census data into 2000 geography.

August 09, 2011

Made with ESRI Business Analyst

©2010 ESRI   www.esri.com/ba   800-447-9778 Try it Now!   Page 10 of 10

5

Determining enterprise zones and other special areas

John Lang, contributor

Objectives

- Show how Table Join and Select by Attributes can be used to determine an area, satisfying given criteria
- Show how Select by Attributes can be used to undertake a "what if?" analysis using ArcGIS software
- Show how ArcGIS tools can be used to determine descriptive statistics of a variable (field)
- Discuss indexes used in economic development and show how indexes can be constructed within ArcGIS
- Exhibit the presentation and report-writing capabilities of ArcGIS

Introduction

The purpose of this chapter is to show the essential role GIS played in obtaining an **enterprise zone** (**EZ**) designation, as defined by California law for the city of San Jose. You will see how a few of the tools described in chapter 1 were applied to determine a proposed EZ. This chapter will highlight the use of Table Join, Select by Attributes, and the Intersect tool. The outline again will be based on the five steps of the Geographic Approach that were discussed in detail in chapter 3: Ask, acquire, examine, analyze, and act.

The Geographic Approach: Ask

The California Enterprise Zone program aims to stimulate development by allowing private sector market forces to revive local economies, specifically by providing tax incentives to businesses.[1] Taxes on some businesses within an approved EZ are reduced, at a cost to the state, not the municipality. The special state tax incentives aim to encourage business investment and promote the creation of new jobs. An EZ designation for a municipality is a helpful tool in promoting local economic development efforts.

Statistical snapshot of San Jose

This sidebar gives a quick statistical snapshot of San Jose, based on recent data from the American Community Survey (ACS). San Jose is the largest of fifteen municipalities in Santa Clara County, containing 204 of the county's 341 census tracts and more than 60 percent of the county's population.

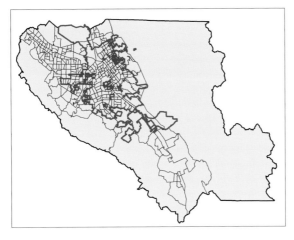

Figure 5S-1. Santa Clara County, its census tracts, and San Jose (purple border). Courtesy of US Census.

Demographics

Today, San Jose is a city of almost a million people. A statistical snapshot of the racial and ethnic makeup of the city indicates the diversity (table 5S-1).

Table 5S-1. Racial and ethnic composition of San Jose

Total	Number	Percent of total population
Total Population	934,415	
White alone	455,770	48.8
Black or African American alone	28,106	3.0
American Indian and Alaska Native alone	5,320	0.6
Asian alone	288,711	30.9
Native Hawaiian and Other Pacific Islander alone	3,369	0.4
Some other race alone	121,358	13.0
Two or more races	31,781	3.4
Hispanic or Latino	294,425	31.5

Courtesy of US Census

Unemployment

San Jose's unemployment rate tends to fluctuate more than the national unemployment rate, partly due to the heavy presence of microprocessor design and manufacturing firms in Silicon Valley. The industry is strongly **procyclical**: national boom times are strong booms in Silicon Valley, and

continued ➡

national downturns are strong downturns in Silicon Valley. More revealing than the unemployment rate at any particular time is the age and gender breakdown of unemployment.

Table 5S-2. Unemployment rates by age and gender in San Jose, 2005–2009

Age	Male	Female
16 to 19 years	28.33	21.22
20 to 21 years	16.05	11.35
22 to 24 years	12.90	9.03
25 to 29 years	8.18	7.09
30 to 34 years	4.99	6.21
35 to 44 years	5.68	6.40
45 to 54 years	6.45	7.33
55 to 59 years	8.05	5.48
60 to 61 years	7.28	7.17
62 to 64 years	7.68	4.84
65 to 69 years	8.00	3.58
70 to 74 years	7.23	8.97
75 and older	9.64	2.29

Courtesy of US Census

Unemployment is heavily concentrated among the young. Generally, the workforce in Silicon Valley is technologically sophisticated compared to the national workforce, even for nontechnological positions.

Poverty and income

Overall, the city poverty rate was just over 10 percent in 2009. Further examination will show that poverty was concentrated in some parts of San Jose and Santa Clara County.

In 2006, twenty-three new EZ designations—each approved for a duration of fifteen years—became available on a competitive basis statewide. The number was limited because of the **tax expenditures** to the state.

Under regulations issued by the California Department of Housing and Community Development pursuant to legislation, selection criteria for EZs focused on areas with low income and high rates of unemployment.[2] While the process also required the city to submit marketing and budget plans, the most critical task for the application was to determine the area or areas that met the basic criteria to be considered for an EZ.

The use of GIS was essential in making this determination. The proposed EZ had to satisfy multiple criteria, some of which might change before the deadline for submission, and the area also had to be geographically contiguous and make economic sense as an EZ.

The minimum requirements for an EZ were stated in terms of characteristics of census tracts. The actual process for reapplying for EZ designation took place in 2006 and used 1990 and 2000 Decennial Census data. The criteria and application in this chapter have been updated to use 2000 census tracts and the most recent three-year ACS data.

The state defined an eligible area as a group of contiguous census tracts, with each tract having a threshold population.[3] Beyond that, the eligible area had to meet three or more indicators of distress. The indicators included requirements like the following:[4]

- The per capita income in each census tract could not exceed a given percentage (say, 80 percent) of the statewide average
- The unemployment rate had to be at least a certain threshold level (roughly 50 percent greater than the national average unemployment rate)
- The percentage of people below the official poverty level had to exceed a threshold level (say, 15 percent)
- The median household income in the tract had to be below a given percent (say, 80 percent) of median household income in the county

In promoting the EZ program, the state is interested in the maximum payoff in terms of ameliorating economic distress for a given tax expenditure. Therefore, the state targets areas that have a combination of distress indicators. Municipalities like San Jose want to propose EZs that meet the state criteria and that make "economic sense" as areas in which to focus economic development efforts. This raises additional considerations, such as the location of "sensitive receptors"—schools, parks, housing for the elderly—and other considerations, such as appropriate transportation infrastructure to support the anticipated economic development. Generally, municipalities would prefer larger rather than smaller EZs because the potential tax revenues depend somewhat on the size of an EZ.

Two central questions will be addressed. First, what areas satisfy these criteria, either one at a time or in various combinations? Second, how do the qualifying areas change when the criteria change? This is the "what if?" question.

To determine the areas that satisfy the criteria, users will rely principally on Table Join and Select by Attributes or the Intersect tool.

The Geographic Approach: Acquire

To identify a suitable area to propose as an EZ, the San Jose Economic Development Department had to rely on several sources of data and to combine data observed at various levels of aggregation. The department confronted two problems that will not be addressed in detail here. The first problem is called the **apportionment problem** (see Kristen S. Kurland and Wilpen L. Gorr, *GIS Tutorial for Health,* Redlands: Esri Press, [2009]). The problem arises when areas don't match. For example, people are often interested in how crime is related to demographic characteristics. Demographic characteristics are easily obtained for census tracts or census block groups. However, crimes are generally reported by police precincts or districts.[5] The police precincts or districts do not generally match census tracts or block groups. A similar issue arises when comparing census variables over time. Census tracts are bureaucratically determined administrative units. They may change from one census to the next. Often, census tracts are split if there has been a sufficient increase in population in an area.

The second problem confronting San Jose economic development officials was that some data they needed were not yet available for the census tract level at that level of disaggregation. The 2006 criteria required recent unemployment data, which were then only available at the county level.

The basic data needed to identify a potential EZ include map files for census tracts and census-tract-level population and economic data (unemployment, income, and poverty status). Because the EZ had to make economic sense as a development area, the following data were also needed:

- Indicators of proportion of land in the area devoted to industrial or commercial development
- Locations of sensitive receptors (schools, centers for the elderly, and so forth)
- Roads and transportation infrastructure

Once the eligible census tracks were identified, the city had to determine a boundary for the EZ. The guidelines allowed for some flexibility in expanding the EZ's boundary beyond the area otherwise deemed eligible under some circumstances. For example, commercial or industrial areas in unqualified census tracts could be included if they were adjacent to qualified census tracts.

Relationships among different geographies

Data are reported for different geographies or levels of aggregation. Some data are individual observations, like the data about biotechnology firms. Other data are totals or averages for geographical areas like counties, municipalities, ZIP Codes, or census tracts. The relationships among these geographical areas are not always straightforward. Census blocks are the lowest level of aggregation for which census data are reported.[6] Collections of blocks are called census block groups. This is the lowest level of aggregation that most readily accessible census data come in. Census block groups consist of one or more census blocks. Census tracts consist of one or more census block groups. Thus, census blocks are always contained in census block groups, which are always contained in census tracts. Census data are also reported for counties and states, based on the census tract information (US Census data can be easily accessed through the US Department of Census American FactFinder website, which features interactive data selection and download).

In addition, census data are reported by place (for example, a municipality), school district, and Core Based Statistical Areas (Metropolitan Statistical Areas and Micropolitan Statistical Areas). Places and school districts may not correspond to counties or states.[7] Figure 5S-2 shows the relationships among the various geographies available in the ACS.

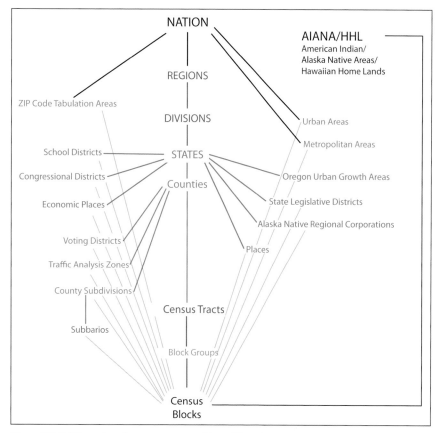

Figure 5S-2. Hierarchy of census geographic entities. Courtesy of US Census.

The Geographic Approach: Examine

The data used to determine an EZ are of varying quality. The data about zoning for San Jose were considered solid. The demographic and income data from the ACS were considered broadly reliable, but because ACS data are based on relatively small samples (approximately 2.5 percent of the population) and margins of error (MOE) are published along with the point estimates (averages), attention should be paid to census tracts where the margin of error was especially large on any of the criteria for qualification for an EZ.

A data quality analysis must include a review of the descriptive statistics of the variables. This can be done in ArcGIS by using the Statistics of a Field functionality (your interface may vary slightly from the ones seen in these examples, depending on the version of ArcGIS used).

It is easy to determine descriptive statistics of any field (variable) in ArcGIS. Opening the attribute table of a layer in a map document also reveals the fields and records behind the map display. The Statistics of a Field function can easily be invoked to show both the central tendency (mean) and dispersion (as shown in a graph of the distribution) of any variable. Within ArcMap, users can open a data table or the attribute table of a layer, highlight a field, and right-click to bring up the context-sensitive menu, and select "Statistics," as shown in figure 5-1.

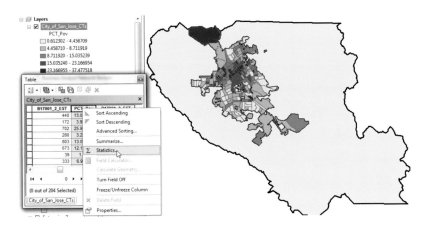

Figure 5-1. Invoking the Statistics of a Field functionality for a map, identifying by census tract the percentage of population living below the federal poverty line in San Jose. Courtesy of US Census.

Statistics of a Field includes statistics such as the count, the minimum, the maximum, the mean (arithmetic average), and the standard deviation (a measure of dispersion). The statistics tool also displays the frequency distribution of the data graphically, via a histogram. For example, the data for percent below the poverty level for the ACS data for the 341 Santa Clara County census tracts in the year 2000 show the mean, standard deviation, and distribution of the Percent in Poverty variable. (Recall from the statistical snapshot that the city-wide poverty rate was just over 10 percent.)

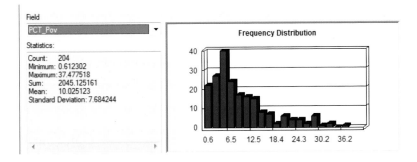

Figure 5-2. Determining the distribution of poverty rates in San Jose using the Field Statistics functionality of ArcGIS. Courtesy of US Census.

Know the data

In undertaking data analysis, the analyst should know the data; that is, fully understand their source, currency, inclusiveness, and other factors before committing to use them in analysis. One might also refer to this as vetting the data. There are several aspects to knowing the data. For example, the analyst should know the following:

- The *source (or sources)* of the data. Because the typical analysis will involve data from several sources, the source of each variable (field) in a dataset should be known.
- The *units of measure* of each variable. For example, "cost" may be a field name. Is this measured in dollars, thousands of dollars, or units of some foreign currency?
- The *time reference* of the data. If the data come from a decennial census, what year was that census—2000, 1990, or 1860?[8] Different variables may have different time references.
- The *aggregation level* of the data. Do the data represent individual transaction data, say, the data about housing prices and floor area mentioned earlier? Such data may be available from a variety of sources, including county assessors' offices. The decennial census also has data on housing prices—"owner-occupier's evaluation of house value"—reported via the American FactFinder website at the census block group and higher level of aggregation. The reported statistic is the *median* for the census block group. Thus, the reported figure is aggregated from individual responses to the block-group level.[9]

Aggregation always destroys information. Analysts can aggregate the data to any desired level if they have the individual transactions data. But if only the aggregate data are available, then the disaggregated data cannot be recovered. The aggregation may *mask variation,* which may lead to false conclusions. The main reason that decennial census data presented on the American FactFinder website are only available at some aggregated level, not as transactions data, is a legal requirement to maintain the privacy of census records for some period of time. Aggregating data is not always a bad thing. While aggregation destroys information by masking variation at the *analysis* stage, disaggregation may mask important conclusions by providing overwhelming detail at the *presentation* stage.

The *number of observations* is often thought to be the most important feature of data—the more observations the better. That is true with one important caveat: assuming the quality of the methodology is the same, the more observations the better. Sometimes, more observations are obtained by loosening controls on sample selection, which can lead to serious problems of biased samples. It is important to know how often the data are updated, if at all. Many of the elements just listed are part of the *metadata* that should be part of every dataset. "Metadata" means roughly "data about the data." It refers to sources and important characteristics of the data. Metadata are provided for many kinds of data, for example, data available through the Esri data portal.

Statistical tools are part of most statistical packages and part of ArcGIS, which help make some parts of knowing the data more routine and enforce a discipline about knowing the data. These statistical tools fall in the general category of "descriptive statistics." The descriptive statistics of a variable or field are basic statistics related to that variable or field. The descriptive statistics are different from the metadata. The metadata refer to the sources and quality of some data; the descriptive statistics refer to the statistical features of the data. The main statistical features of a variable or field are measures of central tendency (averages) and measures of dispersion (how the values are distributed).

Every analysis involving particular variables or fields should begin by determining the descriptive statistics of those variables. This is mainly useful as a check. Are the data reasonable? Does the average of a variable meet expectations, and are the values distributed in an expected way? Perhaps the most important information is also the most basic. The descriptive statistics will reveal the

continued

largest value of the variable (maximum) and the smallest value of the variable (minimum). Many unfortunate errors can be avoided by checking whether these values are reasonable. For example, a problem would become obvious immediately if the descriptive statistics of the income variable indicated that the minimum recorded value was negative! Because income cannot be negative, such a result could indicate a coding error or some other problem with the data.

The Geographic Approach: Analyze

Once the data were obtained, they were brought into the GIS program. The basic cartographic features are indicated in the map in figure 5-3.

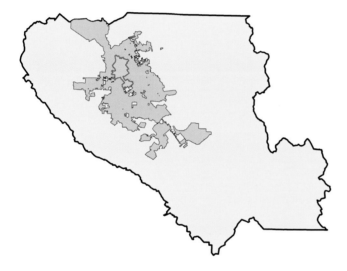

Figure 5-3. Santa Clara County, San Jose (shaded in purple), and the eventual enterprise zone (outlined in red). Courtesy of US Census; City of San Jose, California.

All the elements displayed so far except for the outline of the eventual EZ are map files available for download from the US Census. To do the analysis, these map files need to be joined with the demographic and economic data obtained from the census and other sources. Because census tract-level demographic and economic data and census tract boundary files have a common field, this can be used as a key to join the tabular data to the map. This ArcGIS operation based on a common field is a Table Join, as discussed in chapter 1. Sometimes no common field is available to join different layers. In that case, Spatial Join—a unique capability of a GIS program—can be used to determine how the features relate spatially.

For census map files and tabular data, the common key is often based on a concatenation (stringing together) of the state and county codes and the census tract number. For example, the Table Join used to connect the Santa Clara County census tracts was based on such keys in the map file and in the demographic and economic data file downloaded from the Census Bureau, as shown in figure 5-4.[10]

Once the Table Join is accomplished, the map can be symbolized in a variety of ways. These various symbolizations are easily carried out and changed in a GIS program. Choosing the most appropriate symbolization depends on the purposes of the visualization. As shown in figure 5-5, the most compelling choice for representing different unemployment rates is a color ramp. The figure uses a color ramp to show the distribution of unemployment rates

City_of_San_Jose_CTs

TRACTCE00	CTIDFP00	NAMELSAD00	GEO_ID2	PCT_Pov
505007	06085505007	Census Tract 5050.07	06085505007	13.091342
506601	06085506601	Census Tract 5066.01	06085506601	3.977798
505203	06085505203	Census Tract 5052.03	06085505203	25.961538
511909	06085511909	Census Tract 5119.09	06085511909	3.220242
502300	06085502300	Census Tract 5023	06085502300	13.082437
500100	06085500100	Census Tract 5001	06085500100	12.113031
507002	06085507002	Census Tract 5070.02	06085507002	1.50347
511911	06085511911	Census Tract 5119.11	06085511911	6.940392
512700	06085512700	Census Tract 5127	06085512700	9.572901
504319	06085504319	Census Tract 5043.19	06085504319	5.642458
505700	06085505700	Census Tract 5057	06085505700	14.267887
500300	06085500300	Census Tract 5003	06085500300	28.460851
505800	06085505800	Census Tract 5058	06085505800	4.880582
506203	06085506203	Census Tract 5062.03	06085506203	10.178991
505900	06085505900	Census Tract 5059	06085505900	5.786802

Figure 5-4. The highlighted columns as the basis for making a Table Join. Courtesy of US Census.

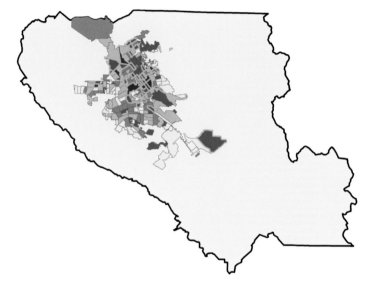

Figure 5-5. Distribution of unemployment rates by census tract in San Jose. Census tracts with unemployment higher than 10 percent are outlined in blue. Courtesy of US Census.

in census tracts in San Jose. The map also shows (highlighted in blue outline) the census tracts with unemployment rates greater than 10 percent.

The map shows that census tracts with high unemployment predominate in some pockets of San Jose. Census tracts with more than 10 percent unemployment were identified by **querying** the database that was created by joining the tabular demographic and economic data to the boundary files.

Using the **Select by Attributes tool**, the census tracts in Santa Clara County that satisfied this criterion can easily be displayed. Figure 5-6 illustrates the **structured query language (SQL)** statement that will identify the census tracts that satisfy the condition.

SQL is a programming language, but it isn't necessary to know SQL to make a selection because the Select by Attributes tool programs the SQL statement for the user. The box at the top of the Select by Attributes dialog box lists all the variables in a layer, as shown in figure 5-6 (your interface may vary slightly, depending on the version of ArcGIS used). The "URate" variable can be identified and selected using the scroll bar.[11] The selection appears in the text box at the bottom of the figure. The focus is on unemployment rates greater than or equal to 10 percent.

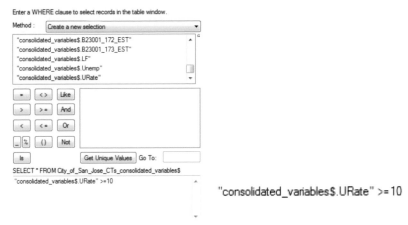

Enter a WHERE clause to select records in the table window.

Method : Create a new selection ▾

"consolidated_variables$.B23001_172_EST"
"consolidated_variables$.B23001_173_EST"
"consolidated_variables$.LF"
"consolidated_variables$.Unemp"
"consolidated_variables$.URate"

=	<>	Like
>	>=	And
<	<=	Or
_%	()	Not
Is		Get Unique Values	Go To:

SELECT * FROM City_of_San_Jose_CTs_consolidated_variables$

"consolidated_variables$.URate" >=10

"consolidated_variables$.URate" >= 10

Figure 5-6. Select by attributes: Census tracts with more than 10 percent unemployment.

Various logical conditions can be constructed, using the keypad on the left-hand side of the middle of the dialog box. The condition ">=" was used and then the number 10 was typed because that is the unemployment rate of interest.

Census tracts can be selected based on satisfying several criteria at once. For example, suppose the focus is on census tracts, in which the median income is $50,000 or less, poverty rates are of at least 10 percent, and unemployment rates are of at least 12 percent. The next step is to construct the SQL statement contained in the Select by Attributes dialog box shown in figure 5-7.

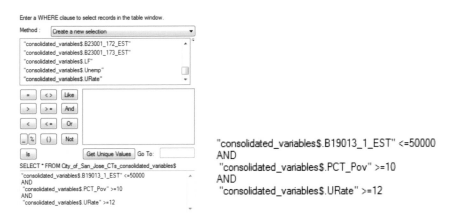

Enter a WHERE clause to select records in the table window.

Method : Create a new selection ▾

"consolidated_variables$.B23001_172_EST"
"consolidated_variables$.B23001_173_EST"
"consolidated_variables$.LF"
"consolidated_variables$.Unemp"
"consolidated_variables$.URate"

=	<>	Like
>	>=	And
<	<=	Or
_%	()	Not
Is		Get Unique Values	Go To:

SELECT * FROM City_of_San_Jose_CTs_consolidated_variables$

"consolidated_variables$.B19013_1_EST" <=50000
AND
"consolidated_variables$.PCT_Pov" >=10
AND
"consolidated_variables$.URate" >=12

"consolidated_variables$.B19013_1_EST" <=50000
AND
"consolidated_variables$.PCT_Pov" >=10
AND
"consolidated_variables$.URate" >=12

Figure 5-7. Selecting multiple attributes.

The result of the selection is the five census tracts highlighted in blue borders in the map in figure 5-8.

The SQL statement in figure 5-7 includes the Boolean logical operator "AND." If the criteria had been stated differently—if, for example, either at least 10 percent of the population had to be below the poverty line, or the median household income had to be below $50,000—users would have used the Boolean operator "OR."[12]

SQL statements, which are easily constructed in the Select by Attributes dialog box in ArcGIS, are the key to undertaking "what if?" analysis in a straightforward way.

GIS can be used to see how each criterion affects the outcome. Figure 5-9 displays the ten census tracts in San Jose that satisfied all the same requirements as the census tracts in the map in figure 5-8, except that the minimum unemployment rate is 10 percent instead of 12 percent.

Literally, all that had to be changed to do the analysis at the new level of unemployment was one number in the SQL statement in the Select by Attributes dialog box.

Figure 5-8. Census tracts (bordered in light blue) that simultaneously feature a low median income, high poverty, and high unemployment in San Jose. Courtesy of US Census.

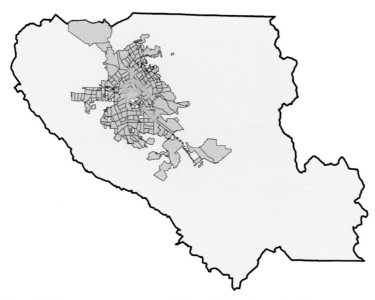

Figure 5-9. The ten census tracts satisfying multiple attributes, including an unemployment rate of at least 10 percent. Courtesy of US Census.

Indexes and economic development analysis

Indexes are essential in economic development analysis because criteria for economic development are stated in terms of indexes, and the success of economic development actions can be determined by the value of indexes. Many indexes used in economic development analysis have a distinct spatial component, explained in detail later in this chapter. ArcGIS provides tools to manipulate indexes and even to create specialty indexes.

What is an index?

An index is a *relative* measure of some value based on a comparison to some reference or base. If the index refers to a phenomenon over time, the base is called a base year or base period. For example, price indexes, which measure the average level of prices over time, are expressed in terms of a base year. Designating a base year creates a base line at that year. Often, indexes are normalized to 100. That is, the value of the index in the base year is set to 100, and index numbers in other years are interpreted as percentage increases or decreases relative to the base year.

Figure 5S-3 shows a map symbolized with an index of real GDP (gross domestic product) per capita (per person) by state.[13] This index has a base year of 2000; the index numbers displayed are for 2010. Index numbers below 100 mean that the real GDP per capita of that state in 2010 was less than GDP of the state in 2000. For example, Georgia has as index number of 93.3. This means that the GDP per person of Georgia in 2010 was just over 93 percent of what it was in the base year, 2000. In other words, Georgia's GDP per person fell about 7 percent from 2000 to 2010. Index numbers above 100 mean that the GDP per person of the state was greater in 2010 than the GDP per person of that state in 2000. North Dakota had an index number of 140.3. This means that the GDP per person of North Dakota in 2010 was more than 40 percent higher than it was in 2000.

Figure 5S-3. Map displaying index numbers of GDP per person by state in 2010 (base year 2000) and identifying by state where the GDP per person declined (light red), was less than the national average (yellow), or greater than the national average (green). Courtesy of US Department of Commerce, Bureau of Economic Development.

continued ➡

The map colors were selected to be reminiscent of a traffic signal.[14] Light red indicates the states in which state GDP per person declined (index values below 100). Yellow indicates states in which index values were more than 100, but less than the national average.[15] Green indicates states in which the growth of GDP per person over the period was greater than the national average.

What are the spatial aspects of indexes?

When index numbers vary over space, a map may tell the story more compellingly than words or numbers. Figure 5S-3 shows an obvious geographic pattern to state real GDP per capita growth in 2010. The worst-performing states were concentrated in the Midwest and South. The Prairie and Rocky Mountain states were the best performing, and the states in the West and Southwest (with the exception of Nevada) all had positive growth in GDP per person.[16]

What can be learned from an index?

Index numbers can powerfully show some phenomena, including relative rates of price inflation. Figure 5S-4 shows the Housing Price Index (published by the Federal Housing Finance Agency) for the United States as a whole and for California. The indexes reveal at a glance something that may not have been evident in the original data. Because the indexes have the same base period (1991–2011-Q1), users can easily compare the growth rates of housing prices in the United States as a whole to those in California.[17]

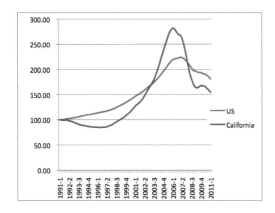

Figure 5S-4. Housing Price Index for the United States and for California (seasonally adjusted) 1991–2011-Q1. Courtesy of Federal Housing Finance Agency.

Rates of growth of housing prices, represented by the slope or steepness of the curves, were greater in California than in the nation as a whole from the late 1990s until about 2005. California housing prices peaked earlier (about 2005) than those for the nation (around 2007), and California housing prices fell more rapidly from their peak than did housing prices in the nation.

Where do indexes come from?

Indexes are published by several governmental agencies and private providers. The Bureau of Labor Statistics publishes perhaps the most widely known index, the Consumer Price Index (CPI). The CPI is used to measure inflation in consumer prices. The index of real state GDP is published by the Bureau of Economic Analysis of the US Commerce Department. The Housing Price Index (HPI) is published by the Federal Housing Finance Agency. The Case-Shiller Home Price Index is published by Standard & Poor's, while the Dow Jones Industrial Average, an index of stock prices, is published by Dow Jones and Company.

continued ➡

Specialty indexes

Economic development analysts may be confronted with a policy issue that is unique to a particular jurisdiction or area. While some indexes involve complex calculations, others can be created within ArcGIS. Indexes involving a single measured element are the most straightforward. An index like the CPI, which involves hundreds of prices, is more complicated because it involves determining the appropriate weight to assign to each price.

This section will describe a situation calling for a specialty index. Index numbers will be computed within ArcGIS; then a fiscal and demographic "what if?" analysis will be conducted involving the newly created index, using ArcGIS. The example will use data for Santa Clara County. A similar analysis could be carried out for any area for which the data were available.

Suppose the county is considering a policy that would subsidize rents in target areas (census tracts) that were deemed "unaffordable."[18] County policymakers are considering a subsidy of a fixed amount, say $1,000, for each rental unit in the target areas. The policy will apply to target areas (census tracts) where the ratio of median rent to median income is more than 30 percent of the countywide average. Policymakers are willing to adjust this standard of affordability, depending on the program's total cost. The program's tentative proposed budget is $5 million annually. Policymakers also want to know the impact of the program on minorities. The task is to identify target areas, then determine the total cost of the subsidy and its impact on minorities.

The basic data for the analysis are the ratio of median rent to median income, the number of rental units, and the proportion of rental units occupied by minority households for census tracts and for the county (base region) as a whole. These data are available from the Census Bureau. For convenience, the data are from the 2000 Decennial Census. While the basic building blocks are available from the census, a special index is needed for analyzing the policy. The next steps are to construct an index, map the result (thus identifying the areas), and undertake a "what if?" analysis using the index.

Building an appropriate index requires a formula to determine the ratios of median income to rent in census tracts relative to median income to rent in the county as a whole:

$$Specialty\ Index = \left[\frac{\frac{median\ rent\ in\ census\ tract}{median\ income\ in\ census\ tract}}{\frac{median\ rent\ in\ county}{median\ income\ in\ county}} \right] * 100$$

The base (denominator) of this index is the ratio of median rent to median income in the county. The index is normalized to 100. That is why 100 appears in the formula for the expression above. If the ratio of rent to median income in a census tract is greater than the countywide average, the index number will be greater than 100. If the ratio of rent to median income in a census tract is less than the countywide average, the index number will be less than 100. To find the census tracts in which the ratio of rent to median income is more than 30 percent, census tracts with index numbers greater than 130 must be identified. If the criterion is changed to apply to areas where the ratio of rent to median income is more than 45 percent of the county average, then census tracts with index numbers greater than 145 would be identified.

The specialty index will be constructed in ArcGIS using a tool that creates a *calculated field*—that is, a field or variable that is not part of the original data but is computed employing a user-specified formula. The formula in the expression will be incorporated into the calculated field.

Census data provide median rents and median income for each census tract and also for the county. The countywide ratio of annualized rents relative to income is 19.1 percent. Taking the ratio of (annualized) median rent to median income in each census tract, dividing that ratio by 19.1, and multiplying the result by 100 will create the index referred to in the expression.

continued

A field called "Index" will be added to the census data for Santa Clara County census tracts, and the field calculator in ArcGIS will be used to incorporate the formula. Next, the Calculator from the Attribute table is invoked, as illustrated in figure 5S-5.

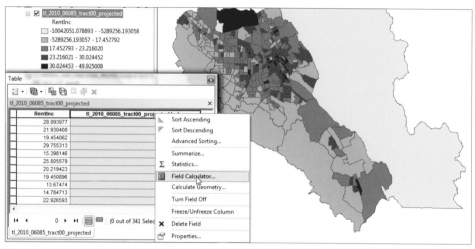

Figure 5S-5. Invoking the Field Calculator for a map of Santa Clara County census tracts. In this case, the color purple was chosen to show the distribution of rent to income. Courtesy of Federal Housing Finance Agency.

The user-specified index formula in the lower text box of the Field Calculator dialog box is incorporated as an SQL statement, as illustrated in figure 5S-6.

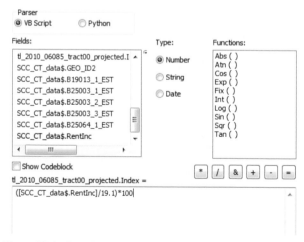

Figure 5S-6. Field Calculator dialog box with the formula for computing the index.

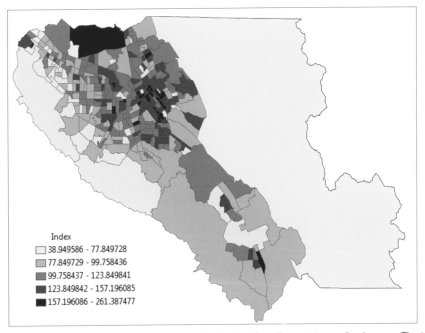

The calculated Index field, which gives the numerical values for the specialty index in terms of the formula above, is shown as the highlighted column in figure 5S-7.

SCC_CTs_Index

NAMELSAD00	Index	RentInc
Census Tract 5052.02	147.088887	28.093977
Census Tract 5050.07	114.818893	21.930408
Census Tract 5066.01	101.853727	19.454062
Census Tract 5052.03	155.786981	29.755313
Census Tract 5053.02	80.618564	15.398146
Census Tract 5065.03	135.107744	25.805579
Census Tract 5091.08	105.860853	20.219423
Census Tract 5091.07	101.837152	19.450896
Census Tract 5099.02	71.5955	13.67474
Census Tract 5100.01	77.406874	14.784713
Census Tract 5087.04	120.034517	22.926593
Census Tract 5119.09	78.622437	15.016886
Census Tract 5023	112.21462	21.432992
Census Tract 5117.01	58.67127	11.206213
Census Tract 5001	146.385155	27.959565

Figure 5S-7. The highlighted Calculated Field "Index" indicates in percentage terms the degree to which the given census tract is over or under the rent-to-income ratio in the county. Courtesy of US Census.

Using the new Index field to symbolize the map, the result is shown in figure 5S-8.

The Selection tool in ArcGIS can be used to see where the ratio in the census tract exceeds the countywide ratio by 30 percent or more, as shown in the map in figure 5S-9.

Index
- 38.949586 - 77.849728
- 77.849729 - 99.758436
- 99.758437 - 123.849841
- 123.849842 - 157.196085
- 157.196086 - 261.387477

Figure 5S-8. Map symbolized using specialized index of median rent to median income. The tan-colored areas correspond to "no data." Courtesy of US Census.

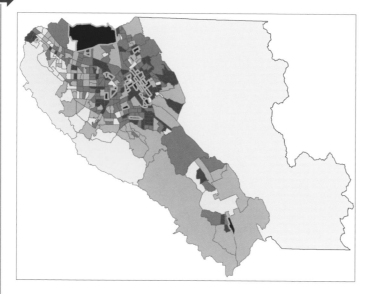

Figure 5S-9. Census tracts highlighted in light blue have ratios of rent to income more than 30 percent of countywide average. Courtesy of US Census.

ArcGIS also can be used to determine the fiscal impact associated with the policy. Generally, determining the fiscal impact of a policy is a complex undertaking because account must be taken of indirect (feedback) effects. This case only accounts for the direct effects of the proposed policy. This analysis requires knowing how many rental housing units are located in the areas identified in the map in figure 5S-8. Invoking the *Statistics of a Field* function gives the mean and the standard deviation of the selected records and also the frequency distribution of rental units. Additionally, the Field Statistics calculator gives the total number (sum) of rental units in the selected census tracts. The alternatives are summarized in table 5S-3.

Table 5S-3. Cost implications of various policy alternatives for subsidizing rents in San Jose

Index value	Number of census tracts for given index value	Number of rental units affected	Cost per unit	Total cost
150	38	29,064	$1,000	$29,064,000
175	15	10,512	$1,000	$10,512,000
185	6	4,929	$1,000	$4,929,000
200	3	2,436	$1,000	$2,436,000

The results show that an affordability standard of 185 (rent-to-income ratio 85 percent greater than the county average) is within the $5 million budgeted for the program.

Sensitivity analysis of a different sort

Sensitivity analysis refers to just the kind of "what if?" questions being asked—how sensitive is the solution to a change in the underlying requirements? But with a problem of the sort represented by EZ requirements, the change in the solution is not just a question of *how much* but of *where*. The spatial element is essential in determining an EZ. The zone must be contiguous and consist of census tracts that are qualified or adjacent to qualified tracts. These spatial elements would be very difficult to manage without a GIS program.

The Geographic Approach: Act

Maps are necessary for the city's application. ArcGIS can generate maps in a variety of formats (as picture files—JPEG, TIFF, GIF—or as PDF files, among others). This output can be used in reports or presentations. In addition, ArcGIS can be used to undertake sophisticated statistical analysis (see chapter 8) with the same data and same format used here. Graphs and reports can be generated in their final form using the Create Graph wizard and Report wizard in ArcGIS.

Summary

In 2007, California approved San Jose's application for an EZ for another fifteen years. GIS played an instrumental role in ultimately determining the size, area, and shape of the San Jose EZ. Meeting the criteria and documenting and justifying the designation of the area were critical to San Jose's success in the competitive process. This chapter and the previous one have mentioned transportation as a significant factor in site selection and the economic viability of special zones. The next chapter examines transportation issues explicitly and the impact of GIS in conjunction with global positioning systems (GPS).

Notes

[1] Enterprise zone regulations, as stated in the California Code of Regulations, Title 25, Subchapter 21, Article 1–13, and statutes, Government Code Sections 7070, et seq.

[2] California Enterprise Zone Program, *Application for Designation—2006*, State of California Business, Transportation and Housing Agency, Department of Housing and Community Development (March 2006).

[3] We have modified the requirements to comport with the more recent data used in this chapter. In the actual 2006 application, census tracts had to have a minimum population of at least 2,500.

[4] We have restated the indicators used in 2006 to cast them in terms of more recent data. The indicators used in 2006 were

1. The net increase in per capita income in the most recent evaluation period was 80 percent or less of the statewide average.

2. The average rate of unemployment for both the beginning and end of the evaluation period was 7.4 percent or more.

3. The percentage of persons below the poverty level at the end of the evaluation period was 15.2 percent or more.

4. At least 70 percent of households had incomes below 80 percent of median county family income at the end of the evaluation period.

5. The area is within a jurisdiction declared a disaster area by the US president within the last seven years. (This last factor was not a consideration in San Jose's application.)

[5] FBI Uniform Crime Reports are available for cities having a population of over 100,000.

[6] See "Census 2010—Census Block Maps" for information about 2010 Census blocks. The census also provides individual data from which information that could identify the individual has been removed. Among the information removed is geographic information. PUMS individual responses are only identified at the PUMA level; see "Public Use Microdata Samples (PUMS)."

[7] For example, the city of Atlanta (Georgia) consists of parts of three counties; the Metropolitan Statistical Area of New York city includes counties in New Jersey.

[8] IPUMS has digitized US Censuses going back to 1850. See "MPC Data Projects," http://www.ipums.umn.edu/.

[9] Individual responses to US census questions are available through Public Use Microdata Samples.

[10] We have retained the original, if uninspired, names of the respective fields: CTIDFP00 in the map file and GEO_ID2 in the economic and demographic data file.

[11] The URate variable is contained in a dataset called "consolidated_variables."

[12] If "OR" were substituted for "AND," 327 of the 341 census tracts in Santa Clara County would qualify. Most do not have the requisite poverty rate or low income, but most do have the minimum population.

[13] Gross domestic product (GDP) is the dollar value of output in the economy, as computed by the Bureau of Economic Analysis of the Department of Commerce. Real GDP controls for the effects of inflation.

[14] This is one kind of diverging color scheme. See "Color Schemes," http://www.fedstats.gov/kids/mapstats/concepts_color.html, "Diverging Color Schemes," http://www.personal.psu.edu/cab38/ColorSch/SchHTMLs/CBColorDiv.html, and "Data Graphics Research," http://geography.uoregon.edu/datagraphics/color_scales.htm for more information.

[15] The national average index value was 106.7 in 2010.

[16] Several Northeast/Atlantic states such as Pennsylvania, West Virginia, and Maryland performed significantly above the national average. It is interesting to note that these states are part of what Paul Krugman identifies as the US Manufacturing Belt (see Paul Krugman, *Geography and Trade* [Cambridge, MA: The MIT Press, 1992], 11–14 and Figure 1.1), which has contained a disproportionate amount of US manufacturing capacity for decades. The persistence of the heavy concentration of manufacturing in this area is explained by the Core-Periphery Model, mentioned in chapter 1, developed by Krugman and others.

[17] The Housing Price Indexes show the changes in median housing price relative to each starting point.

[18] We could cast the same problem in terms of any geographic subunits with reference to a larger base unit. Housing affordability is defined in many different ways. See S. Mathur and A. Parker, *Housing Silicon Valley: A 20-Year Plan to End the Affordable Housing Crisis* (San Francisco, California: Local Initiatives Support Corporation, 2007).

6

Jobs-housing balance, transit-oriented development, and commute time: Integrating GIS and GPS

Objectives

- To discuss the role of global positioning system (GPS) technology in economic development
- To discuss how GPS can be used to assess the traffic and congestion implications of specific economic development projects
- To show how GPS technology can be combined with GIS to enhance economic development analysis

Introduction

Commute distance and travel time affect economic development in profound ways. Accessibility is important in determining the desirability of a site for retail (as seen in chapter 4), manufacturing, and other activity (as seen in chapter 1). Chapter 2 noted that drive times are quite different from "as the crow flies" radii. As with chapter 5, the outline of this chapter will be based on the Geographic Approach: Ask, acquire, examine, analyze, and act.

This chapter examines the economic development aspects of transportation, specifically looking at the intersection of GIS with a related technology, GPS. This chapter discusses two major issues in economic development related to transportation costs and commuting—the jobs-housing balance and transit-oriented development—and also looks at how GPS technology can be used along with GIS by economic development analysts.

GPS has far-ranging applications to economic development analysis—from surveying local businesses to collecting data on commute patterns of workers of a specific employer. In these cases, GPS can be used to collect more accurate data at a lower cost than traditional diary-keeping and survey methods. ArcGIS software tools make analysis of the data easier and provide insights about a variety of topics in economic development, such as site selection and congestion effects. This chapter develops an example from New York City to demonstrate how GPS data can be employed in ArcGIS and combined with other data to understand an individual's **journey-to-work.** The chapter provides several other related examples.

What is GPS?

GPS is a system for identifying a location based on signals from a constellation of twenty-seven satellites orbiting the earth. Reading signals using individual receivers from three or more of these satellites at the same time can pinpoint the exact location of a receiver with latitude and longitude coordinates. A receiver equipped with a memory chip, called a GPS logger, can store tens of thousands of positions for a prescribed time interval, speed, or distance. Some loggers can also record altitude (elevation).

A GPS tracker is a related device for recording the travel itinerary of an individual; rather than storing the traces on a memory chip, the GPS tracker employs a wireless phone line to transmit real-time location information to a central server. Because a GPS tracker requires a separate phone number for each individual—as well as the involvement of private mobile phone companies—GPS trackers are not used in travel surveys as often as GPS loggers but are quite popular in tracking delivery trucks, the movement of children for safety purposes, and transit buses for real-time arrival predictions.[1] GPS loggers and trackers differ from the GPS navigators installed in cars and smart phones. GPS navigators have a screen, maps, and software designed to guide users through traffic to a final destination. GPS navigators usually do not automatically record latitude and longitude information.

Since GPS was first introduced as a component of travel survey data collection in 1996,[2] there have been dozens of GPS-based travel surveys conducted around the world.[3] Most of the earlier GPS travel surveys were vehicle-based and deployed in urban areas where car driving is the predominant mode of transportation. As GPS technology continues to evolve, person-based or handheld GPS data collection is becoming a convenient and cost-effective way to gather information from survey participants. GPS-generated journey-to-work data collection holds great promise, especially in complex urban environments such as Chicago, San Francisco, and New York City, where public transit is a major mode of transportation.

The Geographic Approach: Ask

GPS can be used for many straightforward and useful applications in economic development and other planning situations. For example, handheld GPS units, which record the longitude and latitude of a specific point, are often used to create a dataset of locations—such as locations of fire hydrants—that may not have specific street addresses. Similarly, using GPS may be a more cost-effective way to do a "walking survey" of businesses—locating them with a handheld GPS unit—rather than recording street addresses. These tasks can be done by people with minimal training. The results can be represented in a GIS program, and the tools mentioned in chapter 1 can be used to develop analyses of the data.

Another application of GPS would be to compare alternative routes. Because GPS loggers can record time and position, it is possible to determine speed over different legs of alternative routes. In setting up a delivery route or determining the best route for commuters, information from GPS loggers, combined with GIS, can provide crucial insight. Along the same lines, GPS can be useful in determining the impact on traffic congestion of the siting of a specific new facility.

Journey-to-work and jobs-housing balance

Americans increasingly spend more time commuting to work. The average commute in 1980 was 21.7 minutes. That grew to 22.4 minutes by 1990, and to 25.5 minutes by 2000. More significantly, commute times are becoming more dispersed—the proportion of trips in all categories under twenty minutes declined between 1990 and 2000, while the proportion of trips in all categories twenty-five minutes or more increased between 1990 and 2000.[4] This change in the distribution of commute times has a direct connection with possible explanations for the increase in commute times. Figure 6-1 illustrates this change in the distribution of commute times for the San Francisco Bay Area.

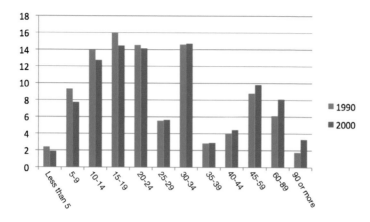

Figure 6-1. Percent of commuters and commute times, San Francisco Bay Area, 1990 and 2000. Courtesy of US Census.

Furthermore, the increase in commute time has both social and economic implications. Longer commutes mean less time with family and lower real earnings for workers. If the wage is constant but the commute time increases, the net wage is lower.

The economic impact of increased commute time is not confined to workers: increased commute time impacts the ability of business firms to attract workers. In a competitive environment, employers would need to pay a "compensating differential" to make up for longer commute times to some facilities. Because the attractiveness of a site depends, among other things, on accessibility to customers or employees, economic development officials have a keen interest in commuting issues.

The issue of jobs-housing balance concerns the proximity of jobs in given pay categories to housing in given price ranges. According to Cervero (1989),[5] part of the reason for the longer journey-to-work trips in the United States is the widening jobs-housing imbalance, a spatial mismatch between the location of jobs and the location of affordable housing.

This imbalance has several causes. It stems, in part, from fiscal and exclusionary zoning policies implemented by many suburban communities that favor commercial and industrial land uses and large lot zoning.[6] The fiscal zoning policies may reduce the number of homes close to employment centers or reduce density, forcing urban development farther from the employment centers, thereby increasing the length of the commute and congestion. The exclusionary zoning policies often drive housing prices upward and make housing less affordable to low-income or even middle-income workers who may like to live near work in the suburban employment centers.

Some policies can mitigate these undesirable effects. Initiatives at the private, local, regional, and state levels, such as inclusionary mixed-use zoning, growth phasing that ties job expansion to housing production, regional tax-sharing programs that remove the fiscal incentive for commercial growth at the expense of residential development, and fair-share housing laws, are recommended by Cervero (1989) to close the jobs-housing gap.

Each of these policies has been employed in various parts of the United States, with some indications of success. Broadly speaking, urban form has a significant effect on **vehicle-miles traveled** (**VMT**), according to Bento et al.[7] Population centrality, city shape, and road density are among factors that can significantly impact household annual VMT.

Journey-to-work has, therefore, become one of the central issues in economic development analysis since human capital constitutes an increasing part of the production costs for goods and services in the information age. Journey-to-work affects human capital in several ways that have not always been appreciated in policy discussions. Longer commutes mean that workers are more tired and distracted on the job, and additional commute time reduces the time available for acquiring new skills. Attention has especially been focused on policy strategies concerning regional jobs-housing balances to shorten the length of the journey-to-work.

The Geographic Approach: Acquire

Journey-to-work is a term used to describe commuting between the place of work and the place of residence of the labor force. Many countries collect journey-to-work data as part of their population censuses. In the United States, the decennial census collects information about where people work, what time they leave for work, how they travel, and how long it takes them to get there.[8] Journey-to-work data are also part of the American Community Survey (ACS). The data available at the census tract level include travel time to work, means of transportation, and time leaving home to go to work. These data can be downloaded from the American Factfinder websites; see the sidebar, "Relationships among different geographies," in chapter 5.

Commuting data and their uses

These journey-to-work data are used by government agencies at all levels for economic, transportation, and environmental planning. For example, these data are used by the US Bureau of Economic Analysis to define economic areas[9] and by the US Census Bureau to define metropolitan statistical areas based on the commuting patterns between and within counties. State, regional, and metropolitan planning organizations (MPOs) use these data to design programs for reducing peak-hour traffic congestion, energy consumption, and the emission of vehicle pollutants. Emergency management officials, mitigation planners, and police and fire departments use the journey-to-work data along with place-of-work data to plan for emergency evacuations in case of disasters.

Measuring imbalance in jobs-housing

While the study of regional jobs-housing balance has gained a great deal of popularity, controversies revolve around how this is to be properly defined and measured. One common measure of jobs-housing balance is the *ratio of jobs to housing units in the area,* with a value of one (and up to 1.5, according to Cervero) representing an area with appropriate balance. However, as Giuliano (1991)[10] notes, a ratio of one may still result in long journey-to-work times if the *mix* of jobs and housing is not compatible. To achieve job-housing balance, the available housing choices should match the earning potential of available jobs in an area. On the other hand, not all commuting to workplaces outside of the area should be considered jobs-housing mismatches. For example, commuting to neighboring areas does not contribute to jobs-housing imbalance if these neighboring areas are within walking distance (the next census tract, for example) and no vehicle traffic is generated (Peng 1997).[11] Another common measure of jobs-housing imbalance is "wasteful commuting,"[12] the difference between the actual commuting and the theoretical minimum commute to connect workers to jobs in a given area. Using journey-to-work data from the 1980 Census, Hamilton (1982, 1989) revealed a very imbalanced urban America, with 90 percent wasteful commuting. White[13] found a much lower amount (11 percent) of wasteful commuting in urban America, using the same data but a different model.

The disparity in these findings provides an example of the importance of properly measuring the jobs-housing balance to derive consistent results. Overwhelmingly, studies pertaining to jobs-housing balance use journey-to-work data at aggregate levels from the census. This represents a problem in terms of the statistical estimation of the jobs-housing imbalance. This is called the modifiable areal unit problem—a statistical bias introduced when data about individuals (such as locations of jobs and housing) are aggregated into areal levels (such as census tracts or counties) for analyses. It causes the results to vary according to the numbers and sizes of the areas used in the statistical analyses. As a result, the larger the areal unit used in a study (for example, census tracts versus census block groups; see the sidebar, "Relationships among different geographies," in chapter 5), the more balanced the area will appear in terms of jobs and housing. Using individualized, scale-invariant journey-to-work data, as supported by GPS, will help to avoid this problem, bring greater consensus to the literature on jobs-housing balance, and make transit policies more effective.[14]

ArcGIS Data Interoperability for Desktop and regional economic development
ArcGIS for Desktop includes an extension called Data Interoperability that allows the software to read data in nonnative formats. These nonnative formats include MapInfo TAB files. They also include many formats in which GPS data may be gathered. The significance of data interoperability for economic development is particularly evident at the regional level. Regional economic development officials may find themselves in the position of having to assemble a dataset from different localities that each uses a different format for maintaining GIS records. This can be a daunting task, which is made easier using ArcGIS Data Interoperability for Desktop and automated geoprocessing workflows (including ModelBuilder), as outlined in chapter 3.

Transit-oriented development

Transit-oriented development is the conscious policy of locating a mix of employment and residences within easy access of transit infrastructure. This is another broad policy designed to alleviate excess commuting. It is hoped that such measures will improve regional mobility and ease transportation constraints on further economic development in urban areas.

To some extent market forces tend to result in firms and people locating near transit infrastructure. The bulk of biotech firms mapped in chapter 1 are concentrated near major highways. Also, the greatest density of population as mapped in chapter 1 is close to major highways.

The Geographic Approach: Examine

Despite the significance of journey-to-work analysis, reliable data at the individual level are difficult to find. Travel surveys are conducted every ten years or so in large cities around the world to obtain individual travel data and to better understand analyses involving journey-to-work. Traditional travel surveys are very time-consuming and costly, requiring participants to fill out many pages of travel diaries on where they have been and at what times they were there during a survey day. Quite often, participants miss some short trips, report trips out of sequence, and approximate departure and arrival times. Furthermore, individual travel surveys may not serve the purposes of a particular business or economic development agency.

Because GPS can accurately record time and location, the technology is supplementing and may eventually replace traditional travel surveys in providing individual journey-to-work data.

How GPS promotes better journey-to-work analysis and commuter policy

GPS can be used to generate highly relevant data, and GIS can be employed to analyze these data. How well and easily do the two technologies work together?

Travel survey data generated using GPS technology can be imported into a GIS program to provide important journey-to-work information such as departure time, arrival time, travel time, workplace location, travel mode, and even trip purpose. Some of these attributes (such as departure and arrival times) are easy to obtain solely from GPS data, while others (such as travel mode and trip purpose) require development of software tools to figure them out. Although attention in the field is still focusing on how to better implement GPS into travel surveys, there have

Data formats

Data come to the economic development official in many formats, as shown in table S6-1.

Table S6-1. Common data formats

Format	Extension	ArcGIS capability
Data Base (dBase) format	.dbf	read directly
Text (ASCII) format	.txt	data interoperability
Microsoft Excel (1997-2003) Workbook	.xls	read directly
Microsoft Excel Workbook	.xlsx	data interoperability
Comma separated values	.csv	data interoperability
Shapefile	.shp	read directly
Digital photograph (joint photographic experts group)	.jpg or .jpeg	read directly
Digital photograph (tagged image file format)	.tiff	read directly
Digital photograph (graphics interchange format)	.gif	read directly
ArcGIS compression format	.sdx	read directly

been some early and promising attempts in developing algorithms and methodologies to extract as much journey-to-work information as possible from the GPS travel survey data.

For example, in one study designed to identify the purpose of 151 vehicle-based trips in the Atlanta metropolitan region, only ten trips (7 percent of the total) were incorrectly assigned a trip purpose.[15] In Toronto, a trip reconstruction software tool was developed to identify four travel modes (walk, bicycle, bus, and car) by travelers in the downtown area. The result was that 92 percent of all trip modes were correctly identified.[16] In New York City, Gong et al. developed a GIS algorithm to identify five travel modes (walk, car, bus, subway, and commuter rail) from person-based GPS travel data. Despite the considerable **urban canyon effect** in Manhattan, 82.6 percent of the trip modes were correctly identified.[17] A GPS receiver identifies its latitude and longitude by locating three or more satellites, calculating the distance to each, and deducing its location through trilateration. Given the speed of the radio signal from the satellite, the distance is determined by the travel time of the signal from the satellite to the receiver. In urban environments with urban canyons created by tall buildings, radio signals may bounce off the surrounding buildings on their way down, causing the receiver to derive longer time and distances than they actually are and therefore identifying an inaccurate location.

The Geographic Approach: Analyze

This section discusses methods used to infer the nature of an individual's daily travel patterns and transportation mode. The example used here of GPS data concerning journey-to-work from New York City will employ data interoperability (see the sidebar on data interoperability in this chapter), Esri Business Analyst Desktop (BA Desktop) (see chapter 2), and additional data.

Importing GPS data into ArcGIS for the New York City example

Two days of journey-to-work data in New York City were recorded by a handheld GPS logger. About the size of a cell phone, the GPS logger was carried in the commuter's pocket (we will call her Mary). The GPS unit was pre-set to record Mary's position every second. For privacy reasons, the home address and name of the commuter have been modified for this example.

The data were generated in .csv format. These files, called **traces**, were imported into a spreadsheet and format-ted for use in ArcGIS. An example of the formatting involved is making sure that the latitude and longitude have the correct sign: positive for latitude (+40.76 degrees) and negative for longitude (-73.83 degrees) to account for Mary's location in the northern and western hemispheres. Once the data were properly formatted in a spreadsheet, the **Add XY Data** tool in ArcGIS was used to import the spreadsheet data and transform the latitude and longitude coordinate pairs into digital points.

Figures 6-2, 6-3a, and 6-3b show the GPS traces for day one and day two of Mary's journey-to-work trips, respectively. Figure 6-2 shows that Mary made three trips on day one. She drove from home to work in the morning (the green stream of points), then went out during the lunch hour to Rego Center for shopping (the purple stream of points), and then drove from work to home in the afternoon (the blue stream of points). (Home and work are shown as circles, and shopping centers are shown as squares.) The green dot represents Mary's home, and the red dot represents her office.

On day two, Mary made two trips, as shown in figures 6-3a and 6-3b. These GPS traces have been cleaned to remove inaccurate GPS points due to the **urban canyon effect**, which tends to cause GPS points to deviate from Mary's travel paths, and also to remove stationary points. First, she took the subway to work in the morning because her car was in a repair shop; the green stream of points recorded her walk to and from subway stations and her ride on an elevated subway train. The stream of points is divided into three segments, as shown in figure 6-3a, because the GPS logger could not receive satellite signals while the subway train was underground. After work Mary took the subway home (the blue stream of points), stopping at the Rego Center (as shown in figure 6-3b).

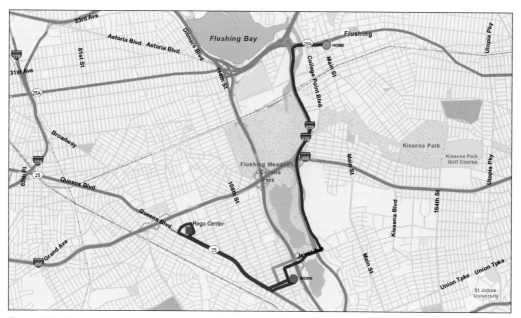

Figure 6-2. Mary's traces for three trips (from home to work, in green; lunchtime shopping at the Rego Center, in purple; and her return from work, in blue) on day one. Data displayed in screenshots of Esri Business Analyst are courtesy of Esri; US Census Bureau; Infogroup; Bureau of Labor Statistics; Applied Geographic Solutions, Inc.; Directory of Major Malls, Inc.; GfK MRI; and Market Planning Solutions, Inc.

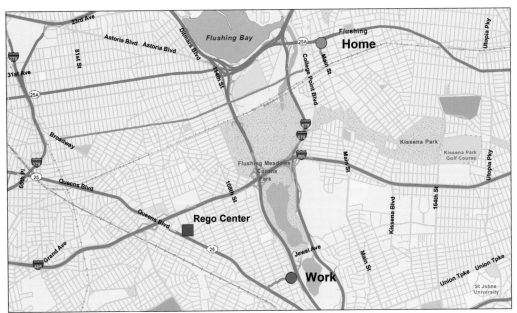

Figure 6-3a. Mary's traces (in green) for the day two trip to work via subway. Data displayed in screenshots of Esri Business Analyst are courtesy of Esri; US Census Bureau; Infogroup; Bureau of Labor Statistics; Applied Geographic Solutions, Inc.; Directory of Major Malls, Inc.; GfK MRI; and Market Planning Solutions, Inc.

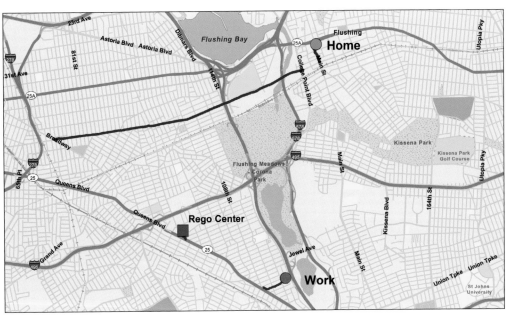

Figure 6-3b. Mary's traces (in blue) for the day two trip home from work via subway with a shopping stop. Data displayed in screenshots of Esri Business Analyst are courtesy of Esri; US Census Bureau; Infogroup; Bureau of Labor Statistics; Applied Geographic Solutions, Inc.; Directory of Major Malls, Inc.; GfK MRI; and Market Planning Solutions, Inc.

Inferring trip purpose from the data

It is important to know why people take trips, known generally as the *trip purpose*. Work trips are very different from nonwork trips. For example, work trips are less flexible in terms of arrival time and destination since employees must arrive on time at the office. The workday is generally eight hours. These are the main reasons for morning and evening rush hours when the amount of vehicle traffic exceeds the capacity of the transportation networks. On the other hand, nonwork trips, such as those for shopping or social visits, are more discretionary in nature: travelers might choose one shopping center over another, travel in the morning or the afternoon, or choose to travel today or wait until tomorrow. Discretionary trips are often made during off-peak hours in order to avoid traffic and are more responsive to transportation policies such as congestion pricing, wherein higher fares are imposed during rush hours for some congested areas of cities.

Unlike time and location information, trip purpose is not inherently an attribute collected by GPS receivers. However, work trips are not difficult to identify from successive days of travel data because a commuter typically repeats the same trip each weekday and often starts and ends these trips around the same time each day. For example, Mary started her morning trips between 7 a.m. and 7:30 a.m. and arrived at her work place before 8 a.m. on both day one and day two.

Compared to traditional, diary-based travel surveys, it is much easier and less costly to use GPS to collect data for multiple days of travel since the survey participants simply need to switch on the GPS logger in the morning and switch it off at night. Most of the traditional paper surveys, by contrast, collected data for just one day due to the tremendous burdens imposed on survey participants since they were asked to write down every trip origin, destination, time period, and travel path over the course of that day.

Even though discretionary trips are more difficult to identify, start and end times of the trip, duration of the trip, and the businesses or land uses at the destination all can be used to infer the trip purpose. Take Mary's second trip on day one as an example (the purple stream of points in figure 6-2). Mary started the trip at 12:10 p.m. and ended the trip at 12:55 p.m., so it can be assumed the trip occurred during her lunch hour. BA Desktop provides a map layer depicting the locations of shopping centers. Adding the shopping center layer to the map

Figure 6-4. Detail of Mary's lunchtime shopping trip (in dark green points) on day one. Data displayed in screenshots of Esri Business Analyst are courtesy of Esri; US Census Bureau; Infogroup; Bureau of Labor Statistics; Applied Geographic Solutions, Inc.; Directory of Major Malls, Inc.; GfK MRI; and Market Planning Solutions, Inc.

reveals a cluster of GPS points near the Rego Center (figure 6-4), indicating that Mary's trip purpose was for shopping. It is important to consider that she may have used the lunch hour either to buy lunch or shop for something, or a combination of both during the forty-five-minute period.

Besides providing business data to help determine why people travel, BA Desktop offers many tools for travel planning. For example, before Mary went out to the Rego Center, she may have used the drive-time trade areas tool to select a shopping center for her shopping trip on day one. Since she is using her lunch hour to do the shopping, Mary wanted to spend no more than twenty minutes traveling, including trip time, parking, and getting to and from parking to work or shopping. This means that the drive-time component of the shopping has to be very short. The Trade Area wizard from the BA Desktop toolbar allows Mary to create two trade areas within three- and eight-minute drive times from her workplace.

Figure 6-5 shows the two trade areas generated from the Drive-time Trade Area tool. The yellow trade area represents areas that can be reached within a three-minute drive-time window, and the orange trade area represents an eight-minute window. Only one shopping center, the Rego Center, falls within the three-minute drive-time trade area. Three other shopping centers, Queens Center, Queens Place, and the Shops at Atlas Park, are reachable within an eight-minute drive time. Mary decided to go to the nearby Rego Center, allowing her more time to pick a birthday present for her husband.

In addition to being useful to Mary for the purpose of shopping for a birthday present, these drive-time areas are useful for economic development analysis. Mary's work location is a major employment center. Businesses that cater to office workers on their lunch breaks (or on their trips to or from work) will be keenly interested in knowing who can reach them within a three- or eight-minute drive. Furthermore, economic development officials will be interested in knowing what kinds of businesses might be attracted to concentrations of employees who can patronize these businesses during lunch breaks or on trips to and from work.

The drive-time polygons have an irregular shape because the travel speed varies with the streets. The polygons stretch out along major streets where travel speeds tend to be higher.

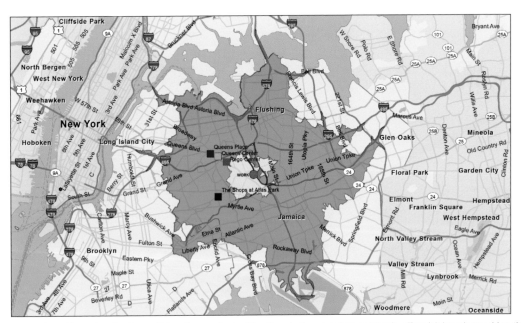

Figure 6-5. Shopping Centers within a three-minute drive time (yellow area) from Mary's office (eight-minute drive-time areas also shown, in orange). Data displayed in screenshots of Esri Business Analyst are courtesy of Esri; US Census Bureau; Infogroup; Bureau of Labor Statistics; Applied Geographic Solutions, Inc.; Directory of Major Malls, Inc.; GfK MRI; and Market Planning Solutions, Inc.

The transport link

Before identifying the travel mode, it's important to determine the transport link upon which an individual travels, whether it is a street, highway, or railroad segment. When a transport link is heavily used and congested, it becomes a traffic bottleneck that lengthens the journey-to-work time and imposes transportation constraints on economic development in urban areas.

Sophisticated techniques match GPS streams to transport links. The basic idea is to find the transport link that most closely coincides with a given GPS stream. A simple example is here. On day two, Mary walked on Seventy-First Avenue from the subway station to her workplace in the morning (the green stream of GPS points at the lower-right corner of figure 6-3a). When viewed in detail, as in figure 6-6, the GPS points appear mostly

Figure 6-6. Mary's trace from the subway station at Queens Boulevard to her work place on 113th Street on day two. Data displayed in screenshots of Esri Business Analyst are courtesy of Esri; US Census Bureau; Infogroup; Bureau of Labor Statistics; Applied Geographic Solutions, Inc.; Directory of Major Malls, Inc.; GfK MRI; and Market Planning Solutions, Inc.

Table — Day2Trip1S3

VALID	LATITUDE	LONGITUDE	SPEEDKMH	HDOP	NSATUSED	NEAR_FID	NEAR_DIST	NEAR_FC
DGPS	40.722197	-73.842289	5.04	1.3	7	53	0.000139	71stAVE
DGPS	40.722203	-73.842277	4.376	0.97	8	53	0.00014	71stAVE
DGPS	40.722214	-73.842259	4.944	1.3	7	53	0.000145	71stAVE
DGPS	40.722213	-73.842244	4.154	1.3	7	53	0.000139	71stAVE
DGPS	40.722218	-73.842224	5.42	1.9	6	53	0.000137	71stAVE
DGPS	40.722226	-73.842199	6.011	1.9	6	53	0.000136	71stAVE
DGPS	40.722234	-73.842174	6.771	1.9	6	53	0.000135	71stAVE
DGPS	40.72224	-73.842145	7.669	1.9	6	53	0.000131	71stAVE
DGPS	40.722251	-73.842122	6.237	1.2	7	53	0.000134	71stAVE
DGPS	40.722251	-73.842111	5.971	1.9	6	53	0.00013	71stAVE
DGPS	40.72226	-73.842101	4.747	1.2	7	53	0.000135	71stAVE
DGPS	40.722269	-73.842087	5.211	1.9	6	53	0.000139	71stAVE
DGPS	40.722274	-73.842065	6.388	1.9	6	53	0.000137	71stAVE

(0 out of 375 Selected)

Day2Trip1S3

Figure 6-7. Output of the Near tool analysis applied to Mary's trace to work on day two. Data displayed in screenshots of Esri Business Analyst are courtesy of Esri; US Census Bureau; Infogroup; Bureau of Labor Statistics; Applied Geographic Solutions, Inc.; Directory of Major Malls, Inc.; GfK MRI; and Market Planning Solutions, Inc.

between Seventieth Avenue and Seventy-First Avenue because of the urban canyon effect. Figure 6-6 shows the GPS points in solid light green dots within the section bounded by Seventieth Avenue, Seventy-First Avenue, Queens Boulevard, and 113th Street.

The GPS points can be matched to a street segment using the **Near tool**—one of the proximity tools in the ArcToolbox. This tool is used in the analysis in chapter 8. Using Near tool analysis, almost all GPS points were matched to Seventy-First Avenue, indicating that Mary traveled from the subway station to her workplace via Seventy-First Avenue. The output of the Near tool is shown in figure 6-7.

Travel mode

The transportation mode that people take to travel to work, whether it is walking, driving, or taking a bus or subway, is an important component of journey-to-work information. Different travel modes reveal their unique patterns in the GPS traces, making it possible for the analyst to use GIS to decipher which mode was used. A number of attributes can be used to help make this determination. First, *speed* provides information about the travel mode. For example, driving is much faster than walking. The average driving speed in New York City is 16.4 miles/hour, or 7.3 meters/second (24 feet/second), while the average walking speed is only 3.5 miles/hour, or 1.6 meters/second (5.2 feet/second). [18]

Figure 6-8 shows Mary's travel near her workplace on day one (the more widely spaced blue points represent driving) and day two (the green points represent walking). In general, the blue points are farther apart than the green points since the GPS logger was set to record locations in one-second intervals; the results reflect the faster travel speeds inherent to driving. Using the **Measure tool** in ArcGIS, it can be determined that the two blue points in the middle of figure 6-8 are approximately 12.2 meters apart, reflecting a travel speed of 12.2 meters/second (39.7 feet/second)—this is much closer to the average driving speed (7.3) than walking speed (1.6) in New York City.

Where GPS traces contain significant lengths of missing data, it can be assumed that the trip included some time spent on an underground train. Driving through a tunnel by car would also cause a brief loss of GPS signals, usually about five to ten minutes, depending on the length of the tunnel, but this signal gap is much shorter than that for the subway mode. Comparing Mary's journey-to-work trip on both days shows that there are no GPS signal gaps on day one (figure 6-2), since Mary drove her car to work, while there are lengthy GPS signal gaps on day two (figures 6-3a and 6-3b) when she took the subway to work.

Mary's trip from her workplace to home on day two can be used as an example to explore this further (figure 6-3b). The reproduced map in figure 6-9 includes selected subway stops (yellow dots). The first stream of blue GPS points at the bottom of the map shows Mary walking from her workplace to the Forest Hills subway station. The

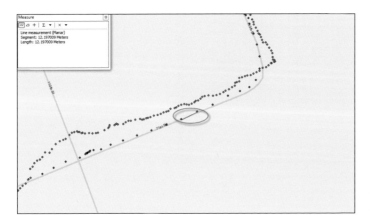

Figure 6-8. Using the Measure tool to determine speed, in this case, the speed of Mary's travels near her workplace. Data displayed in screenshots of Esri Business Analyst are courtesy of Esri; US Census Bureau; Infogroup; Bureau of Labor Statistics; Applied Geographic Solutions, Inc.; Directory of Major Malls, Inc.; GfK MRI; and Market Planning Solutions, Inc.

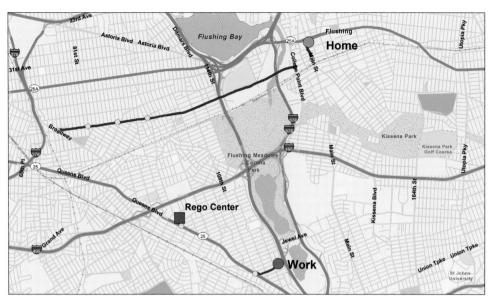

Figure 6-9. Understanding Mary's traces (in blue) for the day two trip from work (red circle) to home (green circle) via subway, with a shopping stop at Rego Center (purple square). Data displayed in screenshots of Esri Business Analyst are courtesy of Esri; US Census Bureau; Infogroup; Bureau of Labor Statistics; Applied Geographic Solutions, Inc.; Directory of Major Malls, Inc.; GfK MRI; and Market Planning Solutions, Inc.

Figure 6-10. Changes in the speed of the subway indicated by the spacing of GPS trace points (in blue). Data displayed in screenshots of Esri Business Analyst are courtesy of Esri; US Census Bureau; Infogroup; Bureau of Labor Statistics; Applied Geographic Solutions, Inc.; Directory of Major Malls, Inc.; GfK MRI; and Market Planning Solutions, Inc.

GPS traces disappear between the Forest Hills station and the Rego Park station (in purple) since Mary was in an underground train and the GPS logger could not establish latitude and longitude coordinates from the satellites. The second stream of GPS points shows a short trip between the Rego Park station and the Rego Center, when Mary got off the subway and went to the shopping center to exchange the birthday present purchased on day one. The GPS signal was lost again between the Rego Park station and the Roosevelt station, shown on the left side of the map, until Mary transferred to a subway line whose path includes elevated tracks above ground level. The GPS logged the third stream of points on the elevated train until it went to the underground Flushing station, showing a brief GPS signal loss right before Mary exited the subway at Flushing station. The last stream of GPS points indicate that Mary walked home from the Flushing station.

GPS points are closer to each other near the stations as the train slows down when approaching the stations and speeds up when departing the stations (as shown in figure 6-10 for selected stations on the elevated portion of Mary's day two return home). These same patterns can be found for above-ground trains (commuter trains or elevated subway trains) or buses in the proximity of stations or stops. This can be differentiated from driving or walking since there are no regular stops with such modes. If a car or pedestrian does stop (for example, to wait for a traffic signal to turn green), this usually happens at street intersections.

The Geographic Approach: Act

This chapter showed through example how GPS can be used to study journey-to-work travels and urban transportation infrastructure for economic development. As GPS and GIS technologies continue to advance, the accuracy of GPS will minimize urban canyon effect, and GIS will offer better tools to derive more travel information from the GPS data. When cell phone towers and wi-fi can be used to supplement satellites to provide location information underground and smart phones equipped with GPS become more widely used, GPS will increasingly become an essential tool for understanding journey-to-work patterns and regional jobs-housing balance. This is evident in the increasing use of smart phones. Adam Smith, in the article "Phone Wars" in *Time* magazine (August 24, 2009), noted that the share of mobile phones that were also smart phones had doubled in three years.

Business Analyst provides many details about the demographic and economic characteristics of drive-time areas, as indicated in the materials in the appendix.

Summary

This chapter examined the intersection of two technologies, GIS and GPS. The combination of GPS and GIS holds great promise for economic development analysis. GPS increases the accuracy and lowers the cost of economic analysis involving data about commuting. GIS makes possible the analysis of these data by economic development officials. Together, these technologies can help economic development officials account for commuting, mitigate the effects of congestion, and limit wasteful commuting.

Appendix

This appendix presents two of the many reports that can be generated by Business Analyst: the Executive Summary report for the three-minute and eight-minute drive-time areas from Mary's work location and the Tapestry Segmentation report for the same drive-time areas.

The Executive Summary gives an overview of the demographic and economic characteristics of the area for the two drive-time areas.

Executive Summary

Drive Time Areas 1		Prepared By Business Analyst Desktop
		Latitude: 40.723126
Drive Time: 3, 8 minutes		Longitude: -73.83817

	0 - 3 minutes	0 - 8 minutes
Population		
1990 Population	92,270	1,289,011
2000 Population	98,895	1,508,134
2010 Population	99,825	1,555,913
2015 Population	100,729	1,580,047
1990-2000 Annual Rate	0.70%	1.58%
2000-2010 Annual Rate	0.09%	0.30%
2010-2015 Annual Rate	0.18%	0.31%
2010 Male Population	47.2%	48.8%
2010 Female Population	52.8%	51.2%
2010 Median Age	42.9	36.3

In the identified market area, the current year population is 1,555,913. In 2000, the Census count in the market area was 1,508,134. The rate of change since 2000 was 0.30 percent annually. The five-year projection for the population in the market area is 1,580,047, representing a change of 0.31 percent annually from 2010 to 2015. Currently, the population is 48.8 percent male and 51.2 percent female.

Population by Employment

Currently, 89.1 percent of the civilian labor force in the indentified market area is employed and 10.9 percent are unemployed. In comparison, 89.2 percent of the U.S. civilian labor force is employed, and 10.8 percent are unemployed. In five years the rate of employment in the market area will be 91.0 percent of the civilian labor force, and unemployment will be 9.0 percent. The percentage of the U.S. civilian labor force that will be employed in five years is 91.2 percent, and 8.8 percent will be unemployed. In 2000, 58.0 percent of the population aged 16 years or older in the market area participated in the labor force, and 0.0 percent were in the Armed Forces.

In the current year, the occupational distribution of the employed population is:

    57.0 percent in white collar jobs (compared to 61.6 percent of the U.S. employment)
    23.9 percent in service jobs (compared to 17.3 percent of U.S. employment)
    19.1 percent in blue collar jobs (compared to 21.1 percent of U.S. employment)

In 2000, 31.0 percent of the market area population drove alone to work, and 1.8 percent worked at home. The average travel time to work in 2000 was 42.5 minutes in the market area, compared to the U.S average of 25.5 minutes.

Population by Education

In the current year, the educational attainment of the population aged 25 years or older in the market area was distributed as follows:

    21.2 percent had not earned a high school diploma (14.8 percent in the U.S)
    30.2 percent were high school graduates only (29.6 percent in the U.S.)
    6.9 percent had completed an Associate degree (7.7 percent in the U.S.)
    18.3 percent had a Bachelor's degree (17.7 percent in the U.S.)
    9.9 percent had earned a Master's/Professional/Doctorate Degree (10.4 percent in the U.S.)

Per Capita Income

1990 Per Capita Income	$23,170	$14,752
2000 Per Capita Income	$29,459	$18,085
2010 Per Capita Income	$35,182	$22,453
2015 Per Capita Income	$41,412	$26,140
1990-2000 Annual Rate	2.43%	2.06%
2000-2010 Annual Rate	1.75%	2.13%
2010-2015 Annual Rate	3.31%	3.09%

Households

1990 Households	44,935	469,683
2000 Households	46,308	513,208
2010 Total Households	46,076	518,350
2015 Total Households	46,275	523,682
1990-2000 Annual Rate	0.30%	0.89%
2000-2010 Annual Rate	-0.05%	0.10%
2010-2015 Annual Rate	0.09%	0.20%
2010 Average Household Size	2.14	2.97

The household count in this market area has changed from 513,208 in 2000 to 518,350 in the current year, a change of 0.10 percent annually. The five-year projection of households is 523,682, a change of 0.20 percent annually from the current year total. Average household size is currently 2.97, compared to 2.91 in the year 2000. The number of families in the current year is 362,121 in the market area.

Data Note: Income is expressed in current dollars
Source: U.S. Bureau and Census, 2000 Census of Population and Housing, ESRI forecast for 2010 and 2015. ESRI converted 1990 Census data into 2000 geography.

August 14, 2011

Made with ESRI Business Analyst

©2010 ESRI   www.esri.com/ba   800-447-9778   Try it Now!   Page 1 of 4

 **esri**

Executive Summary

Drive Time Areas 1

Drive Time: 3, 8 minutes

	0 - 3 minutes	0 - 8 minutes
Households by Income		

Current median household income is $55,507 in the market area, compared to $54,442 for all U.S. households. Median household income is projected to be $65,950 in five years. In 2000, median household income was $41,530.

Current average household income is $66,918 in this market area, compared to $70,173 for all U.S households. Average household income is projected to be $78,331 in five years. In 2000, average household income was $52,522, compared to $40,049 in 1990.

Current per capita income is $22,453 in the market area, compared to the U.S. per capita income of $26,739. The per capita income is projected to be $26,140 in five years. In 2000, the per capita income was $18,085, compared to $14,752 in 1990.

	0 - 3 minutes	0 - 8 minutes
Median Household Income		
2000 Median Household Income	$48,803	$41,530
2010 Median Household Income	$62,068	$55,507
2015 Median Household Income	$73,781	$65,950
2000-2010 Annual Rate	2.37%	2.87%
2010-2015 Annual Rate	3.52%	3.51%
Average Household Income		
1990 Average Household Income	$47,474	$40,049
2000 Average Household Income	$62,762	$52,522
2010 Average Household Income	$76,122	$66,918
2015 Average Household Income	$90,069	$78,331
1990-2000 Annual Rate	2.83%	2.75%
2000-2010 Annual Rate	1.90%	2.39%
2010-2015 Annual Rate	3.42%	3.20%
2010 Housing		
1990 Total Housing Units	47,302	491,415
2000 Total Housing Units	48,016	535,913
2010 Total Housing Units	48,742	552,889
2015 Total Housing Units	49,078	560,199
1990 Owner Occupied Housing Units	16,055	188,490
1990 Renter Occupied Housing Units	28,880	281,193
1990 Vacant Housing Units	2,387	21,748
2000 Owner Occupied Housing Units	17,565	207,037
2000 Renter Occupied Housing Units	28,743	306,171
2000 Vacant Housing Units	1,722	22,732
2010 Owner Occupied Housing Units	17,623	208,958
2010 Renter Occupied Housing Units	28,454	309,392
2010 Vacant Housing Units	2,665	34,539
2015 Owner Occupied Housing Units	17,810	211,967
2015 Renter Occupied Housing Units	28,464	311,715
2015 Vacant Housing Units	2,803	36,517

Currently, 37.8 percent of the 552,889 housing units in the market area are owner occupied; 56.0 percent, renter occupied; and 6.2 are vacant. In 2000, there were 535,913 housing units - 38.6 percent owner occupied, 57.1. percent renter occupied, and 4.2 percent vacant. The rate of change in housing units since 2000 is 0.30 percent. Median home value in the market area is $378,885, compared to a median home value of $157,913 for the U.S. In five years, median value is projected to change by 4.49 percent annually to $472,030. From 2000 to the current year, median home value change by 6.16 percent annually.

Data Note: Income is expressed in current dollars
Source: U.S. Bureau and Census, 2000 Census of Population and Housing, ESRI forecast for 2010 and 2015. ESRI converted 1990 Census data into 2000 geography.

August 14, 2011

Made with ESRI Business Analyst

Prepared By Business Analyst Desktop

	Whole Layer (Drive Time Areas 1)
Population	
1990 Population	1,381,281
2000 Population	1,607,029
2010 Population	1,655,738
2015 Population	1,680,776
1990-2000 Annual Rate	1.53%
2000-2010 Annual Rate	0.29%
2010-2015 Annual Rate	0.30%
2010 Male Population	48.7%
2010 Female Population	51.3%
2010 Median Age	36.7

In the identified market area, the current year population is 1,655,738. In 2000, the Census count in the market area was 1,607,029. The rate of change since 2000 was 0.29 percent annually. The five-year projection for the population in the market area is 1,680,776, representing a change of 0.30 percent annually from 2010 to 2015. Currently, the population is 48.7 percent male and 51.3 percent female.

Population by Employment

Currently, 89.3 percent of the civilian labor force in the indentified market area is employed and 10.7 percent are unemployed. In comparison, 89.2 percent of the U.S. civilian labor force is employed, and 10.8 percent are unemployed. In five years the rate of employment in the market area will be 91.2 percent of the civilian labor force, and unemployment will be 8.8 percent. The percentage of the U.S. civilian labor force that will be employed in five years is 91.2 percent, and 8.8 percent will be unemployed. In 2000, 58.1 percent of the population aged 16 years or older in the market area participated in the labor force, and 0.0 percent were in the Armed Forces.

In the current year, the occupational distribution of the employed population is:

59.5 percent in white collar jobs (compared to 61.6 percent of the U.S. employment)

23.1 percent in service jobs (compared to 17.3 percent of U.S. employment)
18.4 percent in blue collar jobs (compared to 21.1 percent of U.S. employment)

In 2000, 30.7 percent of the market area population drove alone to work, and 1.9 percent worked at home. The average travel time to work in 2000 was 42.4 minutes in the market area, compared to the U.S average of 25.5 minutes.

Population by Education

In the current year, the educational attainment of the population aged 25 years or older in the market area was distributed as follows:

20.3 percent had not earned a high school diploma (14.8 percent in the U.S)

29.5 percent were high school graduates only (29.6 percent in the U.S.)

6.9 percent had completed an Associate degree (7.7 percent in the U.S.)

19.2 percent had a Bachelor's degree (17.7 percent in the U.S.)

10.7 percent had earned a Master's/Professional/Doctorate Degree (10.4 percent in the U.S.)

Per Capita Income	
1990 Per Capita Income	$15,313
2000 Per Capita Income	$18,785
2010 Per Capita Income	$23,220
2015 Per Capita Income	$27,055
1990-2000 Annual Rate	2.06%
2000-2010 Annual Rate	2.09%
2010-2015 Annual Rate	3.10%
Households	
1990 Households	514,618
2000 Households	559,516
2010 Total Households	564,426
2015 Total Households	569,957
1990-2000 Annual Rate	0.84%
2000-2010 Annual Rate	0.09%
2010-2015 Annual Rate	0.20%
2010 Average Household Size	2.91

The household count in this market area has changed from 559,516 in 2000 to 564,426 in the current year, a change of 0.09 percent annually. The five-year projection of households is 569,957, a change of 0.20 percent annually from the current year total. Average household size is currently 2.91, compared to 2.84 in the year 2000. The number of families in the current year is 386,900 in the market area.

Data Note: Income is expressed in current dollars
Source: U.S. Bureau and Census, 2000 Census of Population and Housing, ESRI forecast for 2010 and 2015. ESRI converted 1990 Census data into 2000 geography.

August 14, 2011

Made with ESRI Business Analyst

©2010 ESRI   www.esri.com/ba   800-447-9778 Try it Now!   Page 3 of 4

Whole Layer (Drive Time Areas 1)

Households by Income

Current median household income is $56,064 in the market area, compared to $54,442 for all U.S. households. Median household income is projected to be $66,517 in five years. In 2000, median household income was $42,078.

Current average household income is $67,669 in this market area, compared to $70,173 for all U.S households. Average household income is projected to be $79,284 in five years. In 2000, average household income was $53,368, compared to $40,699 in 1990.

Current per capita income is $23,220 in the market area, compared to the U.S. per capita income of $26,739. The per capita income is projected to be $27,055 in five years. In 2000, the per capita income was $18,785, compared to $15,313 in 1990.

Median Household Income	
2000 Median Household Income	$42,078
2010 Median Household Income	$56,064
2015 Median Household Income	$66,517
2000-2010 Annual Rate	2.84%
2010-2015 Annual Rate	3.48%
Average Household Income	
1990 Average Household Income	$40,699
2000 Average Household Income	$53,368
2010 Average Household Income	$67,669
2015 Average Household Income	$79,284
1990-2000 Annual Rate	2.75%
2000-2010 Annual Rate	2.34%
2010-2015 Annual Rate	3.22%
2010 Housing	
1990 Total Housing Units	538,717
2000 Total Housing Units	583,929
2010 Total Housing Units	601,631
2015 Total Housing Units	609,277
1990 Owner Occupied Housing Units	204,545
1990 Renter Occupied Housing Units	310,074
1990 Vacant Housing Units	24,136
2000 Owner Occupied Housing Units	224,603
2000 Renter Occupied Housing Units	334,913
2000 Vacant Housing Units	24,454
2010 Owner Occupied Housing Units	226,581
2010 Renter Occupied Housing Units	337,845
2010 Vacant Housing Units	37,204
2015 Owner Occupied Housing Units	229,777
2015 Renter Occupied Housing Units	340,179
2015 Vacant Housing Units	39,320

Currently, 37.7 percent of the 601,631 housing units in the market area are owner occupied; 56.2 percent, renter occupied; and 6.2 are vacant. In 2000, there were 583,929 housing units - 38.5 percent owner occupied, 57.4. percent renter occupied, and 4.2 percent vacant. The rate of change in housing units since 2000 is 0.29 percent. Median home value in the market area is $377,232, compared to a median home value of $157,913 for the U.S. In five years, median value is projected to change by 4.56 percent annually to $471,476. From 2000 to the current year, median home value change by 6.20 percent annually.

Data Note: Income is expressed in current dollars
Source: U.S. Bureau and Census, 2000 Census of Population and Housing, ESRI forecast for 2010 and 2015. ESRI converted 1990 Census data into 2000 geography.

August 14, 2011

The Tapestry Segmentation report is based on a categorization of the population of these drive-time areas.

Tapestry Segmentation Area Profile

Drive Time Areas 1	Prepared By Business Analyst Desktop
	Latitude: 40.723126
Drive Time: 3 minutes	Longitude: -73.83817

Top Twenty Tapestry Segments (Tapestry descriptions can be found at: http://www.esri.com/library/whitepapers/pdfs/community-tapestry.pdf)

		Households		U.S. Households		
			Cumulative		Cumulative	
Rank	Tapestry Segment	Percent	Percent	Percent	Percent	Index
1	44. Urban Melting Pot	26.7%	26.7%	0.7%	0.7%	3981
2	30. Retirement Communities	18.1%	44.8%	1.5%	2.1%	1245
3	22. Metropolitans	17.3%	62.0%	1.2%	3.3%	1455
4	23. Trendsetters	11.6%	73.6%	1.1%	4.4%	1100
5	20. City Lights	8.1%	81.8%	1.0%	5.4%	785
	Subtotal	**81.8%**		**5.4%**		
6	11. Pacific Heights	7.7%	89.5%	0.6%	6.0%	1252
7	27. Metro Renters	4.9%	94.4%	1.4%	7.4%	359
8	03. Connoisseurs	3.2%	97.6%	1.4%	8.8%	229
9	05. Wealthy Seaboard Suburbs	1.5%	99.1%	1.4%	10.1%	107
10	08. Laptops and Lattes	0.5%	99.6%	1.0%	11.2%	53
	Subtotal	**17.8%**		**5.8%**		
11	09. Urban Chic	0.4%	100.0%	1.3%	12.5%	31
12	01. Top Rung	0.0%	100.0%	0.7%	13.2%	0
13	02. Suburban Splendor	0.0%	100.0%	1.7%	14.9%	0
14	04. Boomburbs	0.0%	100.0%	2.3%	17.2%	0
15	06. Sophisticated Squires	0.0%	100.0%	2.7%	19.9%	0
	Subtotal	**0.4%**		**8.8%**		
16	07. Exurbanites	0.0%	100.0%	2.5%	22.5%	0
17	10. Pleasant-Ville	0.0%	100.0%	1.7%	24.2%	0
18	12. Up and Coming Families	0.0%	100.0%	3.5%	27.7%	0
19	13. In Style	0.0%	100.0%	2.5%	30.2%	0
20	14. Prosperous Empty Nesters	0.0%	100.0%	1.8%	32.0%	0
	Subtotal	**0.0%**		**12.0%**		
	Total	**100.0%**		**32.0%**		**313**

Top Ten Tapestry Segments Site vs. U.S.

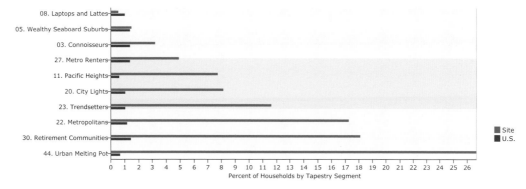

Source: ESRI

August 14, 2011

Made with ESRI Business Analyst

©2010 ESRI   www.esri.com/ba   800-447-9778 Try it Now!   Page 1 of 18

Tapestry Segmentation Area Profile

Drive Time Areas 1

Drive Time: 3 minutes

Tapestry Indexes by Households

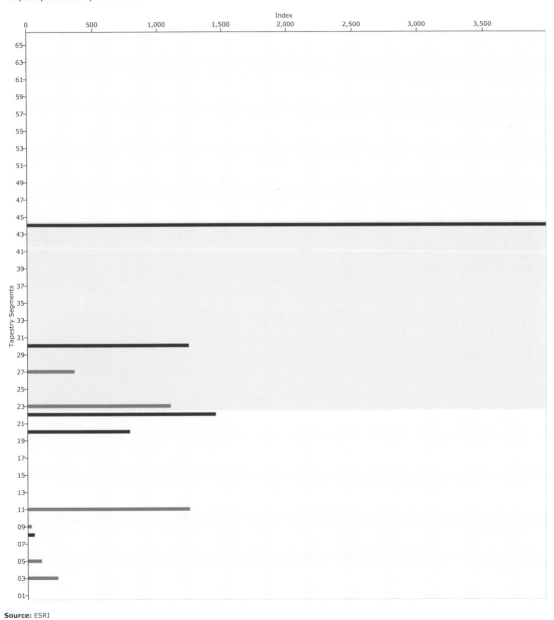

Source: ESRI

August 14, 2011

Made with ESRI Business Analyst

©2010 ESRI   www.esri.com/ba   800-447-9778 Try it Now!   Page 2 of 18

Tapestry Segmentation Area Profile

Drive Time Areas 1

Drive Time: 3 minutes

Tapestry LifeMode Groups	2010 Households		
	Number	Percent	Index
Total:	46,077	100.0%	
L1. High Society	**2,145**	**4.7%**	**37**
01 Top Rung	0	0.0%	0
02 Suburban Splendor	0	0.0%	0
03 Connoisseurs	1,464	3.2%	229
04 Boomburbs	0	0.0%	0
05 Wealthy Seaboard Suburbs	681	1.5%	107
06 Sophisticated Squires	0	0.0%	0
07 Exurbanites	0	0.0%	0
L2. Upscale Avenues	**3,753**	**8.1%**	**59**
09 Urban Chic	190	0.4%	31
10 Pleasant-Ville	0	0.0%	0
11 Pacific Heights	3,563	7.7%	1252
13 In Style	0	0.0%	0
16 Enterprising Professionals	0	0.0%	0
17 Green Acres	0	0.0%	0
18 Cozy and Comfortable	0	0.0%	0
L3. Metropolis	**11,686**	**25.4%**	**484**
20 City Lights	3,736	8.1%	785
22 Metropolitans	7,950	17.3%	1455
45 City Strivers	0	0.0%	0
51 Metro City Edge	0	0.0%	0
54 Urban Rows	0	0.0%	0
62 Modest Income Homes	0	0.0%	0
L4. Solo Acts	**7,861**	**17.1%**	**251**
08 Laptops and Lattes	246	0.5%	53
23 Trendsetters	5,353	11.6%	1100
27 Metro Renters	2,262	4.9%	359
36 Old and Newcomers	0	0.0%	0
39 Young and Restless	0	0.0%	0
L5. Senior Styles	**8,338**	**18.1%**	**147**
14 Prosperous Empty Nesters	0	0.0%	0
15 Silver and Gold	0	0.0%	0
29 Rustbelt Retirees	0	0.0%	0
30 Retirement Communities	8,338	18.1%	1245
43 The Elders	0	0.0%	0
49 Senior Sun Seekers	0	0.0%	0
50 Heartland Communities	0	0.0%	0
57 Simple Living	0	0.0%	0
65 Social Security Set	0	0.0%	0
L6. Scholars & Patriots	**0**	**0.0%**	**0**
40 Military Proximity	0	0.0%	0
55 College Towns	0	0.0%	0
63 Dorms to Diplomas	0	0.0%	0

Source: ESRI

August 14, 2011

Made with ESRI Business Analyst

©2010 ESRI   www.esri.com/ba   800-447-9778   Try it Now!   Page 3 of 18

Tapestry Segmentation Area Profile

Drive Time Areas 1

Drive Time: 3 minutes

Prepared By Business Analyst Desktop
Latitude: 40.723126
Longitude: -73.83817

Tapestry LifeMode Groups	2010 Households		
	Number	Percent	Index
Total:	46,077	100.0%	
L7. High Hopes	**0**	**0.0%**	**0**
28 Aspiring Young Families	0	0.0%	0
48 Great Expectations	0	0.0%	0
L8. Global Roots	**12,294**	**26.7%**	**326**
35 International Marketplace	0	0.0%	0
38 Industrious Urban Fringe	0	0.0%	0
44 Urban Melting Pot	12,294	26.7%	3981
47 Las Casas	0	0.0%	0
52 Inner City Tenants	0	0.0%	0
58 NeWest Residents	0	0.0%	0
60 City Dimensions	0	0.0%	0
61 High Rise Renters	0	0.0%	0
L9. Family Portrait	**0**	**0.0%**	**0**
12 Up and Coming Families	0	0.0%	0
19 Milk and Cookies	0	0.0%	0
21 Urban Villages	0	0.0%	0
59 Southwestern Families	0	0.0%	0
64 City Commons	0	0.0%	0
L10. Traditional Living	**0**	**0.0%**	**0**
24 Main Street, USA	0	0.0%	0
32 Rustbelt Traditions	0	0.0%	0
33 Midlife Junction	0	0.0%	0
34 Family Foundations	0	0.0%	0
L11. Factories & Farms	**0**	**0.0%**	**0**
25 Salt of the Earth	0	0.0%	0
37 Prairie Living	0	0.0%	0
42 Southern Satellites	0	0.0%	0
53 Home Town	0	0.0%	0
56 Rural Bypasses	0	0.0%	0
L12. American Quilt	**0**	**0.0%**	**0**
26 Midland Crowd	0	0.0%	0
31 Rural Resort Dwellers	0	0.0%	0
41 Crossroads	0	0.0%	0
46 Rooted Rural	0	0.0%	0
66 Unclassified	0	0.0%	0

Source: ESRI

Tapestry Segmentation Area Profile

Drive Time Areas 1

Prepared By Business Analyst Desktop
Latitude: 40.723126

Drive Time: 3 minutes

Longitude: -73.83817

Tapestry Urbanization Groups	2010 Households		
	Number	Percent	Index
Total:	46,077	100.0%	
U1. Principal Urban Centers I	**27,454**	**59.6%**	**761**
08 Laptops and Lattes	246	0.5%	53
11 Pacific Heights	3,563	7.7%	1252
20 City Lights	3,736	8.1%	785
21 Urban Villages	0	0.0%	0
23 Trendsetters	5,353	11.6%	1100
27 Metro Renters	2,262	4.9%	359
35 International Marketplace	0	0.0%	0
44 Urban Melting Pot	12,294	26.7%	3981
U2. Principal Urban Centers II	**0**	**0.0%**	**0**
45 City Strivers	0	0.0%	0
47 Las Casas	0	0.0%	0
54 Urban Rows	0	0.0%	0
58 NeWest Residents	0	0.0%	0
61 High Rise Renters	0	0.0%	0
64 City Commons	0	0.0%	0
65 Social Security Set	0	0.0%	0
U3. Metro Cities I	**10,285**	**22.3%**	**197**
01 Top Rung	0	0.0%	0
03 Connoisseurs	1,464	3.2%	229
05 Wealthy Seaboard Suburbs	681	1.5%	107
09 Urban Chic	190	0.4%	31
10 Pleasant-Ville	0	0.0%	0
16 Enterprising Professionals	0	0.0%	0
19 Milk and Cookies	0	0.0%	0
22 Metropolitans	7,950	17.3%	1455
U4. Metro Cities II	**8,338**	**18.1%**	**167**
28 Aspiring Young Families	0	0.0%	0
30 Retirement Communities	8,338	18.1%	1245
34 Family Foundations	0	0.0%	0
36 Old and Newcomers	0	0.0%	0
39 Young and Restless	0	0.0%	0
52 Inner City Tenants	0	0.0%	0
60 City Dimensions	0	0.0%	0
63 Dorms to Diplomas	0	0.0%	0
U5. Urban Outskirts I	**0**	**0.0%**	**0**
04 Boomburbs	0	0.0%	0
24 Main Street, USA	0	0.0%	0
32 Rustbelt Traditions	0	0.0%	0
38 Industrious Urban Fringe	0	0.0%	0
48 Great Expectations	0	0.0%	0

Source: ESRI

August 14, 2011

Made with ESRI Business Analyst

©2010 ESRI

www.esri.com/ba   800-447-9778 Try it Now!

Page 5 of 18

Tapestry Segmentation Area Profile

Drive Time Areas 1

Drive Time: 3 minutes

Prepared By Business Analyst Desktop
Latitude: 40.723126
Longitude: -73.83817

Tapestry Urbanization Groups	2010 Households		
	Number	**Percent**	**Index**
Total:	46,077	100.0%	
U6. Urban Outskirts II	**0**	**0.0%**	**0**
51 Metro City Edge	0	0.0%	0
55 College Towns	0	0.0%	0
57 Simple Living	0	0.0%	0
59 Southwestern Families	0	0.0%	0
62 Modest Income Homes	0	0.0%	0
U7. Suburban Periphery I	**0**	**0.0%**	**0**
02 Suburban Splendor	0	0.0%	0
06 Sophisticated Squires	0	0.0%	0
07 Exurbanites	0	0.0%	0
12 Up and Coming Families	0	0.0%	0
13 In Style	0	0.0%	0
14 Prosperous Empty Nesters	0	0.0%	0
15 Silver and Gold	0	0.0%	0
U8. Suburban Periphery II	**0**	**0.0%**	**0**
18 Cozy and Comfortable	0	0.0%	0
29 Rustbelt Retirees	0	0.0%	0
33 Midlife Junction	0	0.0%	0
40 Military Proximity	0	0.0%	0
43 The Elders	0	0.0%	0
53 Home Town	0	0.0%	0
U9. Small Towns	**0**	**0.0%**	**0**
41 Crossroads	0	0.0%	0
49 Senior Sun Seekers	0	0.0%	0
50 Heartland Communities	0	0.0%	0
U10. Rural I	**0**	**0.0%**	**0**
17 Green Acres	0	0.0%	0
25 Salt of the Earth	0	0.0%	0
26 Midland Crowd	0	0.0%	0
31 Rural Resort Dwellers	0	0.0%	0
U11. Rural II	**0**	**0.0%**	**0**
37 Prairie Living	0	0.0%	0
42 Southern Satellites	0	0.0%	0
46 Rooted Rural	0	0.0%	0
56 Rural Bypasses	0	0.0%	0
66 Unclassified	0	0.0%	0

Data Note: This report identifies neighborhood segments in the area, and describes the socioeconomic quality of the immediate neighborhood. The index is a comparison of the percent of households or population in the area, by Tapestry segment, to the percent of households or population in the United States, by segment. An index of 100 is the US average.

Source: ESRI

August 14, 2011

Tapestry Segmentation Area Profile

Drive Time Areas 1

Prepared By Business Analyst Desktop
Latitude: 40.723126
Longitude: -73.83817

Drive Time: 8 minutes

Top Twenty Tapestry Segments (Tapestry descriptions can be found at: http://www.esri.com/library/whitepapers/pdfs/community-tapestry.pdf)

| | | Households | | U.S. Households | | |
| | | Percent | Cumulative Percent | Percent | Cumulative Percent | Index |
Rank	Tapestry Segment					
1	44. Urban Melting Pot	32.7%	32.7%	0.7%	0.7%	4882
2	20. City Lights	18.4%	51.1%	1.0%	1.7%	1776
3	35. International Marketplace	14.1%	65.2%	1.3%	3.0%	1090
4	45. City Strivers	7.0%	72.2%	0.7%	3.7%	946
5	11. Pacific Heights	6.2%	78.4%	0.6%	4.4%	999
	Subtotal	**78.4%**		**4.4%**		
6	30. Retirement Communities	3.7%	82.1%	1.5%	5.8%	255
7	21. Urban Villages	2.9%	85.1%	0.8%	6.6%	381
8	61. High Rise Renters	2.9%	88.0%	0.7%	7.3%	441
9	22. Metropolitans	2.0%	90.0%	1.2%	8.4%	169
10	23. Trendsetters	2.0%	92.0%	1.1%	9.5%	186
	Subtotal	**13.6%**		**5.1%**		
11	05. Wealthy Seaboard Suburbs	1.4%	93.4%	1.4%	10.9%	101
12	47. Las Casas	1.0%	94.4%	0.8%	11.6%	138
13	34. Family Foundations	1.0%	95.4%	0.8%	12.5%	116
14	10. Pleasant-Ville	0.9%	96.2%	1.7%	14.2%	52
15	03. Connoisseurs	0.9%	97.1%	1.4%	15.6%	62
	Subtotal	**5.1%**		**6.1%**		
16	29. Rustbelt Retirees	0.6%	97.7%	2.1%	17.6%	30
17	58. NeWest Residents	0.5%	98.2%	0.9%	18.5%	58
18	27. Metro Renters	0.4%	98.7%	1.4%	19.9%	32
19	36. Old and Newcomers	0.4%	99.0%	1.9%	21.8%	19
20	65. Social Security Set	0.3%	99.3%	0.6%	22.5%	41
	Subtotal	**2.2%**		**6.9%**		
	Total	**99.3%**		**22.5%**		**442**

Top Ten Tapestry Segments Site vs. U.S.

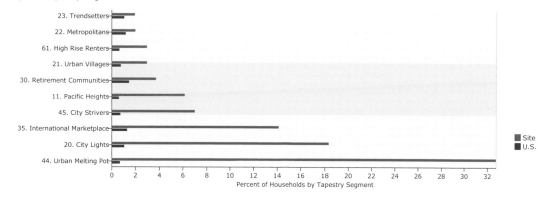

Source: ESRI

August 14, 2011

Made with ESRI Business Analyst

©2010 ESRI

www.esri.com/ba   800-447-9778 Try it Now!

Page 7 of 18

Tapestry Segmentation Area Profile

Drive Time Areas 1

Drive Time: 8 minutes

Prepared By Business Analyst Desktop
Latitude: 40.723126
Longitude: -73.83817

Tapestry Indexes by Households

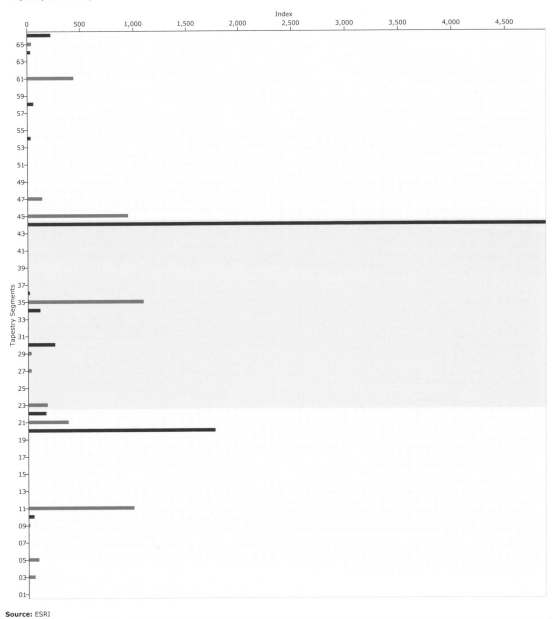

Source: ESRI

Tapestry Segmentation Area Profile

Drive Time Areas 1   Prepared By Business Analyst Desktop
                                                                         Latitude: 40.723126
Drive Time: 8 minutes   Longitude: -73.83817

Tapestry LifeMode Groups	2010 Households		
	Number	Percent	Index
Total:	518,349	100.0%	
L1. High Society	**11,633**	**2.2%**	**18**
01 Top Rung	0	0.0%	0
02 Suburban Splendor	0	0.0%	0
03 Connoisseurs	4,427	0.9%	62
04 Boomburbs	0	0.0%	0
05 Wealthy Seaboard Suburbs	7,206	1.4%	101
06 Sophisticated Squires	0	0.0%	0
07 Exurbanites	0	0.0%	0
L2. Upscale Avenues	**37,349**	**7.2%**	**52**
09 Urban Chic	819	0.2%	12
10 Pleasant-Ville	4,557	0.9%	52
11 Pacific Heights	31,973	6.2%	999
13 In Style	0	0.0%	0
16 Enterprising Professionals	0	0.0%	0
17 Green Acres	0	0.0%	0
18 Cozy and Comfortable	0	0.0%	0
L3. Metropolis	**142,448**	**27.5%**	**524**
20 City Lights	95,144	18.4%	1776
22 Metropolitans	10,364	2.0%	169
45 City Strivers	36,381	7.0%	946
51 Metro City Edge	0	0.0%	0
54 Urban Rows	559	0.1%	31
62 Modest Income Homes	0	0.0%	0
L4. Solo Acts	**14,553**	**2.8%**	**41**
08 Laptops and Lattes	246	0.0%	5
23 Trendsetters	10,178	2.0%	186
27 Metro Renters	2,262	0.4%	32
36 Old and Newcomers	1,867	0.4%	19
39 Young and Restless	0	0.0%	0
L5. Senior Styles	**24,163**	**4.7%**	**38**
14 Prosperous Empty Nesters	332	0.1%	3
15 Silver and Gold	0	0.0%	0
29 Rustbelt Retirees	3,232	0.6%	30
30 Retirement Communities	19,233	3.7%	255
43 The Elders	0	0.0%	0
49 Senior Sun Seekers	0	0.0%	0
50 Heartland Communities	0	0.0%	0
57 Simple Living	0	0.0%	0
65 Social Security Set	1,366	0.3%	41
L6. Scholars & Patriots	**0**	**0.0%**	**0**
40 Military Proximity	0	0.0%	0
55 College Towns	0	0.0%	0
63 Dorms to Diplomas	0	0.0%	0

Source: ESRI

Part II: Applying the Geographic Approach and GIS to economic development analysis

Drive Time Areas 1

Prepared By Business Analyst Desktop
Latitude: 40.723126
Longitude: -73.83817

Drive Time: 8 minutes

Tapestry LifeMode Groups	2010 Households		
	Number	Percent	Index
Total:	518,349	100.0%	
L7. High Hopes	**0**	**0.0%**	**0**
28 Aspiring Young Families	0	0.0%	0
48 Great Expectations	0	0.0%	0
L8. Global Roots	**266,186**	**51.4%**	**627**
35 International Marketplace	73,314	14.1%	1090
38 Industrious Urban Fringe	0	0.0%	0
44 Urban Melting Pot	169,589	32.7%	4882
47 Las Casas	5,388	1.0%	138
52 Inner City Tenants	0	0.0%	0
58 NeWest Residents	2,659	0.5%	58
60 City Dimensions	0	0.0%	0
61 High Rise Renters	15,236	2.9%	441
L9. Family Portrait	**16,405**	**3.2%**	**40**
12 Up and Coming Families	0	0.0%	0
19 Milk and Cookies	0	0.0%	0
21 Urban Villages	15,274	2.9%	381
59 Southwestern Families	0	0.0%	0
64 City Commons	1,131	0.2%	32
L10. Traditional Living	**5,587**	**1.1%**	**12**
24 Main Street, USA	515	0.1%	4
32 Rustbelt Traditions	0	0.0%	0
33 Midlife Junction	0	0.0%	0
34 Family Foundations	5,072	1.0%	116
L11. Factories & Farms	**0**	**0.0%**	**0**
25 Salt of the Earth	0	0.0%	0
37 Prairie Living	0	0.0%	0
42 Southern Satellites	0	0.0%	0
53 Home Town	0	0.0%	0
56 Rural Bypasses	0	0.0%	0
L12. American Quilt	**0**	**0.0%**	**0**
26 Midland Crowd	0	0.0%	0
31 Rural Resort Dwellers	0	0.0%	0
41 Crossroads	0	0.0%	0
46 Rooted Rural	0	0.0%	0
66 Unclassified	25	0.0%	224

Source: ESRI

August 14, 2011

Made with ESRI Business Analyst
©2010 ESRI   www.esri.com/ba   800-447-9778 Try it Now!

Tapestry Segmentation Area Profile

Drive Time Areas 1

Drive Time: 8 minutes

Prepared By Business Analyst Desktop
Latitude: 40.723126
Longitude: -73.83817

Tapestry Urbanization Groups	2010 Households		
	Number	**Percent**	**Index**
Total:	518,349	100.0%	
U1. Principal Urban Centers I	**397,980**	**76.8%**	**981**
08 Laptops and Lattes	246	0.0%	5
11 Pacific Heights	31,973	6.2%	999
20 City Lights	95,144	18.4%	1776
21 Urban Villages	15,274	2.9%	381
23 Trendsetters	10,178	2.0%	186
27 Metro Renters	2,262	0.4%	32
35 International Marketplace	73,314	14.1%	1090
44 Urban Melting Pot	169,589	32.7%	4882
U2. Principal Urban Centers II	**62,720**	**12.1%**	**256**
45 City Strivers	36,381	7.0%	946
47 Las Casas	5,388	1.0%	138
54 Urban Rows	559	0.1%	31
58 NeWest Residents	2,659	0.5%	58
61 High Rise Renters	15,236	2.9%	441
64 City Commons	1,131	0.2%	32
65 Social Security Set	1,366	0.3%	41
U3. Metro Cities I	**27,373**	**5.3%**	**47**
01 Top Rung	0	0.0%	0
03 Connoisseurs	4,427	0.9%	62
05 Wealthy Seaboard Suburbs	7,206	1.4%	101
09 Urban Chic	819	0.2%	12
10 Pleasant-Ville	4,557	0.9%	52
16 Enterprising Professionals	0	0.0%	0
19 Milk and Cookies	0	0.0%	0
22 Metropolitans	10,364	2.0%	169
U4. Metro Cities II	**26,172**	**5.0%**	**47**
28 Aspiring Young Families	0	0.0%	0
30 Retirement Communities	19,233	3.7%	255
34 Family Foundations	5,072	1.0%	116
36 Old and Newcomers	1,867	0.4%	19
39 Young and Restless	0	0.0%	0
52 Inner City Tenants	0	0.0%	0
60 City Dimensions	0	0.0%	0
63 Dorms to Diplomas	0	0.0%	0
U5. Urban Outskirts I	**515**	**0.1%**	**1**
04 Boomburbs	0	0.0%	0
24 Main Street, USA	515	0.1%	4
32 Rustbelt Traditions	0	0.0%	0
38 Industrious Urban Fringe	0	0.0%	0
48 Great Expectations	0	0.0%	0

Source: ESRI

August 14, 2011

Made with ESRI Business Analyst

©2010 ESRI   www.esri.com/ba   800-447-9778 Try it Now!

Page 11 of 18

Tapestry Segmentation Area Profile

Drive Time Areas 1

Prepared By Business Analyst Desktop

Latitude: 40.723126

Drive Time: 8 minutes

Longitude: -73.83817

Tapestry Urbanization Groups	2010 Households		
	Number	**Percent**	**Index**
Total:	518,349	100.0%	
U6. Urban Outskirts II	**0**	**0.0%**	**0**
51 Metro City Edge	0	0.0%	0
55 College Towns	0	0.0%	0
57 Simple Living	0	0.0%	0
59 Southwestern Families	0	0.0%	0
62 Modest Income Homes	0	0.0%	0
U7. Suburban Periphery I	**332**	**0.1%**	**0**
02 Suburban Splendor	0	0.0%	0
06 Sophisticated Squires	0	0.0%	0
07 Exurbanites	0	0.0%	0
12 Up and Coming Families	0	0.0%	0
13 In Style	0	0.0%	0
14 Prosperous Empty Nesters	332	0.1%	3
15 Silver and Gold	0	0.0%	0
U8. Suburban Periphery II	**3,232**	**0.6%**	**6**
18 Cozy and Comfortable	0	0.0%	0
29 Rustbelt Retirees	3,232	0.6%	30
33 Midlife Junction	0	0.0%	0
40 Military Proximity	0	0.0%	0
43 The Elders	0	0.0%	0
53 Home Town	0	0.0%	0
U9. Small Towns	**0**	**0.0%**	**0**
41 Crossroads	0	0.0%	0
49 Senior Sun Seekers	0	0.0%	0
50 Heartland Communities	0	0.0%	0
U10. Rural I	**0**	**0.0%**	**0**
17 Green Acres	0	0.0%	0
25 Salt of the Earth	0	0.0%	0
26 Midland Crowd	0	0.0%	0
31 Rural Resort Dwellers	0	0.0%	0
U11. Rural II	**0**	**0.0%**	**0**
37 Prairie Living	0	0.0%	0
42 Southern Satellites	0	0.0%	0
46 Rooted Rural	0	0.0%	0
56 Rural Bypasses	0	0.0%	0
66 Unclassified	25	0.0%	224

Data Note: This report identifies neighborhood segments in the area, and describes the socioeconomic quality of the immediate neighborhood. The index is a comparison of the percent of households or population in the area, by Tapestry segment, to the percent of households or population in the United States, by segment. An index of 100 is the US average.

Source: ESRI

Tapestry Segmentation Area Profile

Whole Layer (Drive Time Areas 1)   Prepared By Business Analyst Desktop

Top Twenty Tapestry Segments (Tapestry descriptions can be found at: http://www.esri.com/library/whitepapers/pdfs/community-tapestry.pdf)

| | | Households | | U.S. Households | | |
| | | | Cumulative | | Cumulative | |
Rank	Tapestry Segment	Percent	Percent	Percent	Percent	Index
1	44. Urban Melting Pot	32.2%	32.2%	0.7%	0.7%	4808
2	20. City Lights	17.5%	49.7%	1.0%	1.7%	1695
3	35. International Marketplace	13.0%	62.7%	1.3%	3.0%	1001
4	45. City Strivers	6.4%	69.2%	0.7%	3.7%	868
5	11. Pacific Heights	6.3%	75.5%	0.6%	4.4%	1019
	Subtotal	**75.5%**		**4.4%**		
6	30. Retirement Communities	4.9%	80.4%	1.5%	5.8%	336
7	22. Metropolitans	3.2%	83.6%	1.2%	7.0%	274
8	23. Trendsetters	2.8%	86.4%	1.1%	8.1%	260
9	21. Urban Villages	2.7%	89.1%	0.8%	8.8%	350
10	61. High Rise Renters	2.7%	91.8%	0.7%	9.5%	405
	Subtotal	**16.3%**		**5.1%**		
11	05. Wealthy Seaboard Suburbs	1.4%	93.2%	1.4%	10.9%	101
12	03. Connoisseurs	1.0%	94.2%	1.4%	12.3%	75
13	47. Las Casas	1.0%	95.2%	0.8%	13.0%	126
14	34. Family Foundations	0.9%	96.1%	0.8%	13.9%	107
15	10. Pleasant-Ville	0.8%	96.9%	1.7%	15.6%	48
	Subtotal	**5.1%**		**6.1%**		
16	27. Metro Renters	0.8%	97.7%	1.4%	16.9%	59
17	29. Rustbelt Retirees	0.6%	98.2%	2.1%	19.0%	28
18	58. NeWest Residents	0.5%	98.7%	0.9%	19.9%	53
19	36. Old and Newcomers	0.3%	99.0%	1.9%	21.8%	17
20	65. Social Security Set	0.2%	99.3%	0.6%	22.5%	37
	Subtotal	**2.4%**		**6.9%**		
	Total	**99.3%**		**22.5%**		**442**

Top Ten Tapestry Segments Site vs. U.S.

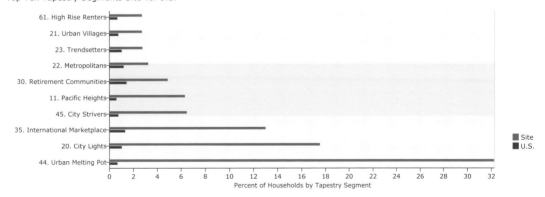

Source: ESRI

August 14, 2011

Made with ESRI Business Analyst

©2010 ESRI   www.esri.com/ba   800-447-9778 Try it Now!   Page 13 of 18

Part II: Applying the Geographic Approach and GIS to economic development analysis

Tapestry Indexes by Households

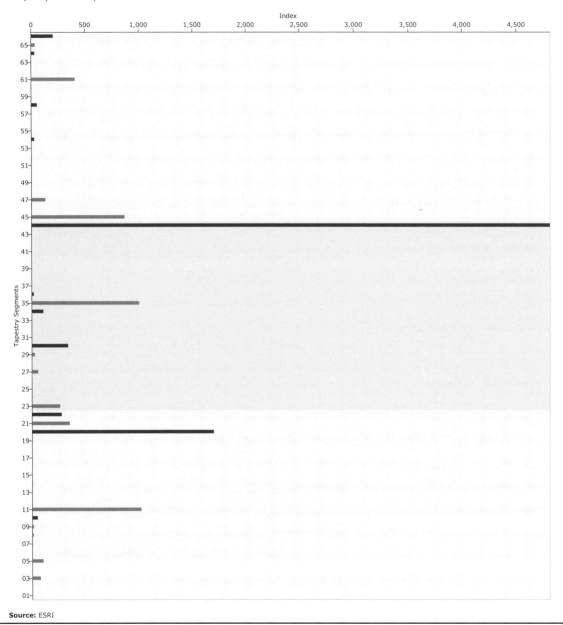

Source: ESRI

Tapestry LifeMode Groups	2010 Households		
	Number	Percent	Index
Total:	564,426	100.0%	
L1. High Society	**13,778**	**2.4%**	**19**
01 Top Rung	0	0.0%	0
02 Suburban Splendor	0	0.0%	0
03 Connoisseurs	5,891	1.0%	75
04 Boomburbs	0	0.0%	0
05 Wealthy Seaboard Suburbs	7,887	1.4%	101
06 Sophisticated Squires	0	0.0%	0
07 Exurbanites	0	0.0%	0
L2. Upscale Avenues	**41,102**	**7.3%**	**53**
09 Urban Chic	1,009	0.2%	13
10 Pleasant-Ville	4,557	0.8%	48
11 Pacific Heights	35,536	6.3%	1019
13 In Style	0	0.0%	0
16 Enterprising Professionals	0	0.0%	0
17 Green Acres	0	0.0%	0
18 Cozy and Comfortable	0	0.0%	0
L3. Metropolis	**154,133**	**27.3%**	**521**
20 City Lights	98,879	17.5%	1695
22 Metropolitans	18,314	3.2%	274
45 City Strivers	36,381	6.4%	868
51 Metro City Edge	0	0.0%	0
54 Urban Rows	559	0.1%	29
62 Modest Income Homes	0	0.0%	0
L4. Solo Acts	**22,414**	**4.0%**	**58**
08 Laptops and Lattes	492	0.1%	9
23 Trendsetters	15,531	2.8%	260
27 Metro Renters	4,524	0.8%	59
36 Old and Newcomers	1,867	0.3%	17
39 Young and Restless	0	0.0%	0
L5. Senior Styles	**32,501**	**5.8%**	**47**
14 Prosperous Empty Nesters	332	0.1%	3
15 Silver and Gold	0	0.0%	0
29 Rustbelt Retirees	3,232	0.6%	28
30 Retirement Communities	27,571	4.9%	336
43 The Elders	0	0.0%	0
49 Senior Sun Seekers	0	0.0%	0
50 Heartland Communities	0	0.0%	0
57 Simple Living	0	0.0%	0
65 Social Security Set	1,366	0.2%	37
L6. Scholars & Patriots	**0**	**0.0%**	**0**
40 Military Proximity	0	0.0%	0
55 College Towns	0	0.0%	0
63 Dorms to Diplomas	0	0.0%	0

Source: ESRI

August 14, 2011

Made with ESRI Business Analyst

©2010 ESRI   www.esri.com/ba   800-447-9778 Try it Now!   Page 15 of 18

Tapestry LifeMode Groups	2010 Households		
	Number	Percent	Index
Total:	564,426	100.0%	
L7. High Hopes	**0**	**0.0%**	**0**
28 Aspiring Young Families	0	0.0%	0
48 Great Expectations	0	0.0%	0
L8. Global Roots	**278,481**	**49.3%**	**603**
35 International Marketplace	73,314	13.0%	1001
38 Industrious Urban Fringe	0	0.0%	0
44 Urban Melting Pot	181,884	32.2%	4808
47 Las Casas	5,388	1.0%	126
52 Inner City Tenants	0	0.0%	0
58 NeWest Residents	2,659	0.5%	53
60 City Dimensions	0	0.0%	0
61 High Rise Renters	15,236	2.7%	405
L9. Family Portrait	**16,405**	**2.9%**	**37**
12 Up and Coming Families	0	0.0%	0
19 Milk and Cookies	0	0.0%	0
21 Urban Villages	15,274	2.7%	350
59 Southwestern Families	0	0.0%	0
64 City Commons	1,131	0.2%	30
L10. Traditional Living	**5,587**	**1.0%**	**11**
24 Main Street, USA	515	0.1%	4
32 Rustbelt Traditions	0	0.0%	0
33 Midlife Junction	0	0.0%	0
34 Family Foundations	5,072	0.9%	107
L11. Factories & Farms	**0**	**0.0%**	**0**
25 Salt of the Earth	0	0.0%	0
37 Prairie Living	0	0.0%	0
42 Southern Satellites	0	0.0%	0
53 Home Town	0	0.0%	0
56 Rural Bypasses	0	0.0%	0
L12. American Quilt	**0**	**0.0%**	**0**
26 Midland Crowd	0	0.0%	0
31 Rural Resort Dwellers	0	0.0%	0
41 Crossroads	0	0.0%	0
46 Rooted Rural	0	0.0%	0
66 Unclassified	25	0.0%	206

Source: ESRI

August 14, 2011

Tapestry Segmentation Area Profile

Whole Layer (Drive Time Areas 1)   Prepared By Business Analyst Desktop

Tapestry Urbanization Groups	2010 Households		
	Number	**Percent**	**Index**
Total:	564,426	100.0%	
U1. Principal Urban Centers I	**425,434**	**75.4%**	**963**
08 Laptops and Lattes	492	0.1%	9
11 Pacific Heights	35,536	6.3%	1019
20 City Lights	98,879	17.5%	1695
21 Urban Villages	15,274	2.7%	350
23 Trendsetters	15,531	2.8%	260
27 Metro Renters	4,524	0.8%	59
35 International Marketplace	73,314	13.0%	1001
44 Urban Melting Pot	181,884	32.2%	4808
U2. Principal Urban Centers II	**62,720**	**11.1%**	**235**
45 City Strivers	36,381	6.4%	868
47 Las Casas	5,388	1.0%	126
54 Urban Rows	559	0.1%	29
58 NeWest Residents	2,659	0.5%	53
61 High Rise Renters	15,236	2.7%	405
64 City Commons	1,131	0.2%	30
65 Social Security Set	1,366	0.2%	37
U3. Metro Cities I	**37,658**	**6.7%**	**59**
01 Top Rung	0	0.0%	0
03 Connoisseurs	5,891	1.0%	75
05 Wealthy Seaboard Suburbs	7,887	1.4%	101
09 Urban Chic	1,009	0.2%	13
10 Pleasant-Ville	4,557	0.8%	48
16 Enterprising Professionals	0	0.0%	0
19 Milk and Cookies	0	0.0%	0
22 Metropolitans	18,314	3.2%	274
U4. Metro Cities II	**34,510**	**6.1%**	**56**
28 Aspiring Young Families	0	0.0%	0
30 Retirement Communities	27,571	4.9%	336
34 Family Foundations	5,072	0.9%	107
36 Old and Newcomers	1,867	0.3%	17
39 Young and Restless	0	0.0%	0
52 Inner City Tenants	0	0.0%	0
60 City Dimensions	0	0.0%	0
63 Dorms to Diplomas	0	0.0%	0
U5. Urban Outskirts I	**515**	**0.1%**	**1**
04 Boomburbs	0	0.0%	0
24 Main Street, USA	515	0.1%	4
32 Rustbelt Traditions	0	0.0%	0
38 Industrious Urban Fringe	0	0.0%	0
48 Great Expectations	0	0.0%	0

Source: ESRI

Part II: Applying the Geographic Approach and GIS to economic development analysis

Chapter 6: Jobs-housing balance, transit-oriented development, and commute time: Integrating GIS and GPS   149

Tapestry Urbanization Groups	2010 Households		
	Number	**Percent**	**Index**
Total:	564,426	100.0%	
U6. Urban Outskirts II	**0**	**0.0%**	**0**
51 Metro City Edge	0	0.0%	0
55 College Towns	0	0.0%	0
57 Simple Living	0	0.0%	0
59 Southwestern Families	0	0.0%	0
62 Modest Income Homes	0	0.0%	0
U7. Suburban Periphery I	**332**	**0.1%**	**0**
02 Suburban Splendor	0	0.0%	0
06 Sophisticated Squires	0	0.0%	0
07 Exurbanites	0	0.0%	0
12 Up and Coming Families	0	0.0%	0
13 In Style	0	0.0%	0
14 Prosperous Empty Nesters	332	0.1%	3
15 Silver and Gold	0	0.0%	0
U8. Suburban Periphery II	**3,232**	**0.6%**	**6**
18 Cozy and Comfortable	0	0.0%	0
29 Rustbelt Retirees	3,232	0.6%	28
33 Midlife Junction	0	0.0%	0
40 Military Proximity	0	0.0%	0
43 The Elders	0	0.0%	0
53 Home Town	0	0.0%	0
U9. Small Towns	**0**	**0.0%**	**0**
41 Crossroads	0	0.0%	0
49 Senior Sun Seekers	0	0.0%	0
50 Heartland Communities	0	0.0%	0
U10. Rural I	**0**	**0.0%**	**0**
17 Green Acres	0	0.0%	0
25 Salt of the Earth	0	0.0%	0
26 Midland Crowd	0	0.0%	0
31 Rural Resort Dwellers	0	0.0%	0
U11. Rural II	**0**	**0.0%**	**0**
37 Prairie Living	0	0.0%	0
42 Southern Satellites	0	0.0%	0
46 Rooted Rural	0	0.0%	0
56 Rural Bypasses	0	0.0%	0
66 Unclassified	25	0.0%	206

Data Note: This report identifies neighborhood segments in the area, and describes the socioeconomic quality of the immediate neighborhood. The index is a comparison of the percent of households or population in the area, by Tapestry segment, to the percent of households or population in the United States, by segment. An index of 100 is the US average.

Source: ESRI

Notes

1 See http://www.nextbus.com for an example.

2 P. Stopher and S. P. Greavers, "Housing Travel Surveys: Where are we going?" *Transportation Research Part A: Policy and Practice* v. 41 (2007): 367–381.

3 The paper by Catherine T. Lawson, Cynthia Chen, Hongmian Gong, Sowmya Karthikeyan, and Alain Kornhauser, "GPS Pilot Project—Phase One: Literature and Product Review," prepared for New York Metropolitan Transportation Council (2007).

4 Clara Reschovsky, "Journey to Work: 2000," *Census 2000 Brief* (March 2004): 5.

5 Robert Cervero, "Jobs-Housing Balancing and Regional Mobility," *Journal of American Planning Association* 55(2) (Spring 1989): 136–150.

6 See J. M. Pogodzinski and Tim Sass, "The Theory and Estimation of Endogenous Zoning," *Regional Science and Urban Economics* 24 (1994): 601–630.

7 Antonio M. Bento, Maureen L. Cropper, Ahmed Mushfiq Mobarak, and Katja Vinha, "The Effects of Urban Spatial Structure on Travel Demand in the United States," *The Review of Economics and Statistics* 87(3) (August 2005): 466–478.

8 Clara Reschovsky, *Census 2000 Brief* (March 2004): 5, US Department of Commerce, Economics and Statistics Administration, Washington, DC.

9 Kenneth P. Johnson and Lyle Spatz, "BEA Economic Areas: A Progress Report on Redefinition—Bureau of Economic Analysis," *Survey of Current Business* (November 1993), 77–79.

10 Genevieve Giuliano, "Is Jobs-Housing Balance a Transportation Issue?" *Transportation Research Record*, no. 1305 (1991): 305–312.

11 Zhong-Ren Peng, "The Jobs-Housing Balance and Urban Commuting," *Urban Studies* 34(8) (1997): 1215–1235.

12 Bruce W Hamilton, "Wasteful Commuting," *Journal of Political Economy* 90(5) (1982); "Wasteful Commuting Again," Journal of Political Economy 97(6) (1989): 1497–1504.

13 Michelle J. White, "Urban Commuting Journeys are not 'Wasteful,'" *Journal of Political Economy* 96(5) (1988): 1097–1110.

14 Mark W. Horner and Bernadette M. Marion, "A Spatial Dissimilarity-Based Index of the Jobs Housing Balance: Conceptual Framework and Empirical Tests," *Urban Studies* 46(3) (2009): 499–517.

15 Jean Wolf, Randall Guensler, and William Bachman, "Elimination of the Travel Diary: An Experiment to Derive Trip Purpose from GPS Travel Data," *Transportation Research Record* no. 1768 (2001): 125–134.

16 Eui-Hwan Chung and Amer Shalaby, "A Trip Reconstruction Tool for GPS-Based Personal Travel Surveys," *Transportation Planning and Technology* 28(5) (2005): 381–401.

17 Hongmian Gong, Cynthia Chen, Evan Bialostozky, and Catherine T. Lawson, "A GPS/GIS Method for Travel Mode Detection in New York City," *Computers, Environment and Urban System* (2011). doi:10.1016/j.compenvurbsys.2011.05.003.

18 Parsons Brinckerhoff Quade & Douglas, Inc., and NuStats International, "General Final Report for the RT-HIS: Regional Travel—Household Interview Survey," prepared for the New York Metropolitan Transportation Council and the North Jersey Transportation Planning Authority, Inc. (February 2000).

Part III
Further explorations in economic
development and GIS

7

Geocoding locations and applications for economic development

Objectives

- To define geocoding and describe the typical inputs and outputs of the process
- To illustrate examples of geocoding locations for economic development purposes
- To distinguish between the geocoding capabilities of Esri Business Analyst Online and Esri Business Analyst Desktop
- To explore uses of geocoded points for cluster analysis as well as transportation cost studies, using ArcGIS Network Analyst

The concepts of *location* and *proximity* form the basis of all geographic analysis. This chapter will reinforce these two most fundamental uses of GIS: to place the locations of real-world features on maps and to study the relative proximity of these features to other features, thus revealing spatial patterns of significance—a central component of all economic development analyses. GIS users commonly convert tabular location information into points on maps. They might do this with a database of company addresses, a mailing list of customer addresses, ZIP Codes possessing certain demographic characteristics, and latitude/longitude coordinates of locations obtained by using a GPS device. The geoprocess of creating "pin maps" is called geocoding.

Chapter 1 briefly explored the results of geocoding by presenting the locations of biotech firms in the San Francisco area that were derived from a spreadsheet of addresses provided by the BayBio trade organization. This chapter draws a distinction between the ease of using Business Analyst Online (BAO) geocoding tools and the full control over the geocoding process offered in Business Analyst Desktop (BA Desktop) and ArcGIS software. The chapter then explores how geocoded location data can be used for more advanced analysis involving transportation access using ArcGIS Network Analyst. The chapter also touches on the use of ArcGIS to create maps for analyzing the clustering of geocoded points, setting the stage for chapter 8, in which GIS tools are used to help determine if the patterns of clustering "hot spots" are statistically significant or merely random.

Geocoding and economic development

Economic development practitioners require the highest level of accurate locational information available. With accurate geocoding information in hand, certain fundamental questions can be addressed, for example:

- What areas of a community were missed or overlooked when searching for potential clients?
- Is there a strong relationship between sales regions and client customers?
- What geographic area is currently served?
- Where are the store locations abundant in a particular region, which might suggest underserved markets elsewhere?

The ability to exploit address and other locational information is essential in carrying out economic development analysis. Since economic development projects usually have large address lists, this chapter will focus on the use of ArcGIS software and Business Analyst to convert these lists into map layers that reveal the spatial distribution of the input address data. The map in figure 7-1 shows geocoded data from an example used in chapter 3; the red-orange dots show the locations of coffeehouses in San Francisco. The addresses were compiled by the authors from an assortment of coffeehouse company websites, pasted into a spreadsheet, and converted via the geocoding process into the points you see below.

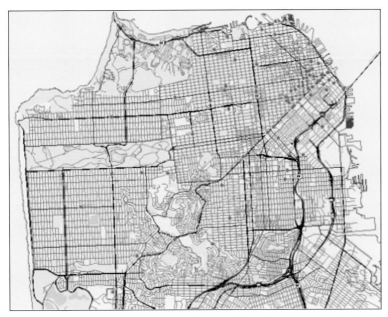

Figure 7-1. Thematic map of geocoded coffeehouses in San Francisco, represented by red-orange dots. The spatial distribution of these business (and customer) locations has numerous applications to economic development projects. Courtesy of City and County of San Francisco Enterprise GIS

With such a map in hand, spatially rooted questions of interest to an economic development practitioner can be examined, such as "Which neighborhood has the highest concentration of coffeehouses, and which do not, suggesting a market opportunity?" Or, perhaps after adding income distribution to the map, "How many coffeehouses are located within a quarter-mile distance of the most affluent parts of the city?"

Economic development analysis wants to account for several factors associated with location—industry clustering, transportation cost, commuting cost, and distance to institutions of higher education, for example. These features need to be located in relation to one another to complete the analysis. Recall that the analysis in chapter 1 located biotechnology firms, research and comprehensive universities, and community colleges, then determined their proximity to such sites as the target population of potential students of community college programs aimed at increasing the number of qualified biotechnology workers. While some data were contained in census files, such as demographic information, other data involved geocoding locations based on an address or other positional information, such as latitude/longitude coordinates.

In comparing the results of a geocoding operation to a spreadsheet list of addresses, such a list might represent, for example, a company's distribution centers or addresses of customer survey respondents. Discerning spatial patterns from this list alone is nearly impossible, of course. Before the advent of GIS software, an economic development professional would most likely tack a paper community map to the wall and push pins into the map at locations of interest. Through the process of geocoding using ArcGIS, BAO, or BA Desktop, the possibilities for geospatial analysis inherent in any address list are revealed once the results of the process produce a digital pin map of plotted addresses. Through a simple visual inspection or by using the set of spatial analysis tools available in ArcGIS, an analyst can determine, for example, the distribution of customers across a given market area, the proximity of households to an important store location, the number of addresses within a given unit of geography (such as census tracts or municipalities), and strategies for the most efficient delivery routing needed to serve all of the mapped addresses.

Using Business Analyst Online for geocoding and proximity analysis

Economic development analysts now have powerful tools to make geocoding a relatively straightforward process. BAO makes geocoding simple by leveraging constantly updated, built-in location data. A later section of this chapter will examine how geocoding can be customized in BA Desktop and ArcGIS using *address locators* along with geographic data supplied by local municipalities. Once again, the point is to draw a distinction between "quick ease of use" and "full control" that the various GIS tools have to offer, so this chapter includes many images from these tools to illustrate these traits.

This chapter will use the following hypothetical scenario to explore the applications of geocoding using BAO: San Jose State University (SJSU) is considering a policy that would require freshmen to live on campus if they graduated from a high school more than thirty miles away. The proposed policy also comes at a time when the university is assertively seeking higher enrollment of African Americans and Hispanics as part of its core mission. SJSU

Figure 7-2. BAO allows users to enter locations in many ways, including street addresses, latitude/longitude coordinate pairs, drawing a polygon on the map, and importing a file of up to one hundred addresses. Here, the address of the university was entered manually. Data displayed in screenshots of Esri Business Analyst are courtesy of Esri; US Census Bureau; Infogroup; Bureau of Labor Statistics; Applied Geographic Solutions, Inc.; Directory of Major Malls, Inc.; GfK MRI; and Market Planning Solutions, Inc.

is aware that creating a distance requirement for dorm living could impact low-income areas—where many freshmen are minorities—by forcing them to pay for housing. The university has tasked the project team and its leader to determine the income and racial characteristics of the region captured within the thirty-mile "as the crow flies" area, as well as a forty-mile area for comparison. Since college students are not crows and travel road networks to reach campus, the project team will compare straight-line distance areas to a forty-five-minute "drive-time" area that accounts for travel along the region's streets. The rationale is that while the university can easily explain a thirty-mile straight-line distance, it isn't reflective of the actual routes that student commuters use to reach campus. The analysis results need to show the race and income characteristics of each of the two distance-based areas to determine if the university's proposed policy disproportionately impacts low-income students.

The project team begins the analysis with BAO to geocode the location of SJSU. Recall from chapter 2 that after logging into the BAO website, users are then prompted to select a location (your interface may vary slightly, depending on the version of the software you are using). Figure 7-2 shows the street address for the university.

BAO is used to geocode the university instantly, as shown in figure 7-3.

The project team then can add rings (also known as buffers) from the geocoded point representing the university's location using the hyperlinks inside of the box associated with the displayed point. Entering thirty miles and forty miles yields the map in figure 7-4.

The team next instructs BAO to delineate a forty-five-minute drive-time area, yielding the result shown in purple in figure 7-5.

Figure 7-3. BAO generates geocoded points and also provides the opportunity to expand the analysis by creating rings (buffers) or drive times based on the geocoded location (SJSU) as a starting point. Data displayed in screenshots of Esri Business Analyst are courtesy of Esri; US Census Bureau; Infogroup; Bureau of Labor Statistics; Applied Geographic Solutions, Inc.; Directory of Major Malls, Inc.; GfK MRI; and Market Planning Solutions, Inc.

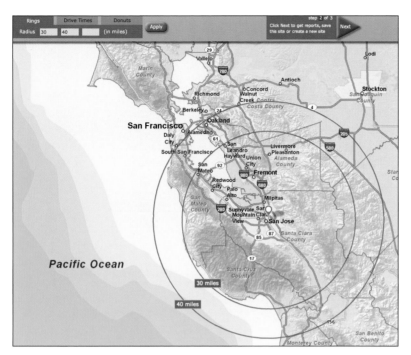

Figure 7-4. BAO can generate straight-line distances from geocoded points (SJSU in this case), within which summary statistics can be generated using BAO's built-in report functionality and demographic and market data. Data displayed in screenshots of Esri Business Analyst are courtesy of Esri; US Census Bureau; Infogroup; Bureau of Labor Statistics; Applied Geographic Solutions, Inc.; Directory of Major Malls, Inc.; GfK MRI; and Market Planning Solutions, Inc.

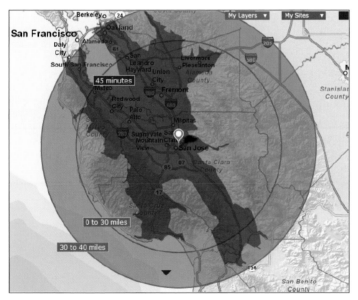

Figure 7-5. BAO can generate straight-line distances as well as drive-time distances (shown here in purple) that are oddly shaped because they reflect the actual road network. Note that the large unpopulated areas east of the university are captured in the straight-line distance areas but correctly absent in the drive-time areas (shown in orange and green). Data displayed in screenshots of Esri Business Analyst are courtesy of Esri; US Census Bureau; Infogroup; Bureau of Labor Statistics; Applied Geographic Solutions, Inc.; Directory of Major Malls, Inc.; GfK MRI; and Market Planning Solutions, Inc.

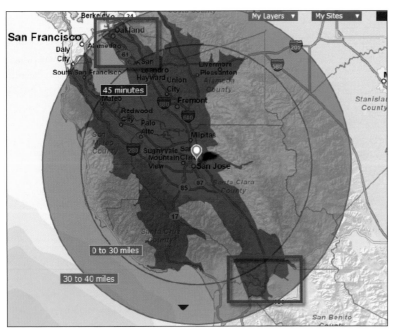

Figure 7-6. A comparison of straight-line distance areas (concentric green and red circles) to drive-time distance areas (purple and blue polygons), as measured from the SJSU campus (blue pin in center). Note the portions of the drive-time distance areas highlighted with red rectangles; these areas fall within the forty-five-minute drive time but beyond the thirty-mile straight-line distance. Different spatial analysis tools (straight-line versus drive-time) yield significantly different results, which in turn prompts discussions about implications for proposed policies. Data displayed in screenshots of Esri Business Analyst are courtesy of Esri; US Census Bureau; Infogroup; Bureau of Labor Statistics; Applied Geographic Solutions, Inc.; Directory of Major Malls, Inc.; GfK MRI; and Market Planning Solutions, Inc.

A closer examination of the map shows that the forty-mile radius encompasses almost exactly the forty-five-minute drive-time area, but the thirty-mile radius cuts off some of those areas (as shown in figure 7-6, highlighted with red rectangles), while including others just within the forty-five-minute drive time.

Some of the areas shown in red have a higher proportion of lower income and African American and Hispanic residents than average. The BAO report-generating tools helped the team determine the median household income and number of African Americans and Hispanics in all three study areas: the straight-line thirty-mile and thirty-to forty-mile radii, and the forty-five-minute drive-time area.

Table 7-1. BAO generated summary data for the three study areas

Distance	Median household income*	Number African-American**	Number Hispanic*
Thirty-mile radius	$93,451	113,828	844,710
Thirty- to forty-mile radius	$75,884	166,776	282,325
Forty-five minute drive-time areas	$87,520	276,396	992,795
The single asterisks indicate that the data reflect the latest census figures. The double-asterisk indicates that the data reflect the 2000 census.			

Table 7-1 suggests that by substituting a forty-mile straight-line radius for a thirty-mile straight-line radius, the "dorm required" policy would apply to areas comprising an additional 167,000 African Americans and 228,000 Hispanics. The area captured within the forty-mile radius also would include areas whose annual median household income is about $75,884, compared to $93,451 for those living within the thirty-mile radius. Now, extending the analysis to include a forty-five-minute drive time (that reflects actual, on-the-ground travel paths) shows that about 276,000 African Americans (rather than approximately 114,000) would be encompassed by the "dorm required" policy. Similarly, 993,000 Hispanics (rather than the approximately 845,000) would be encompassed by the same policy. On average, the forty-five-minute drive-time area has a lower median household income of approximately $87,500, compared with the approximately $93,000 household income for the thirty-mile straight-line distance (dorm-required area) sought by the university.

With these maps and summary data in hand, the project team might recommend that the university use drive time instead of the proposed "as the crow flies" requirement because the latter disproportionately impacts minorities and people of lower income.

Using Business Analyst Desktop for geocoding

The previous example reflected how easy it is to use BAO to generate maps, distance buffers, and summary data. The chapter now examines how BA Desktop handles geocoding. BA Desktop can be used to help integrate all of the powerful ArcGIS spatial analysis tools, such as geoprocessing, into the analysis. The primary similarity between the two Business Analyst versions is the large quantity of reference data and pre-built tools that are available to the user, with little setup required.

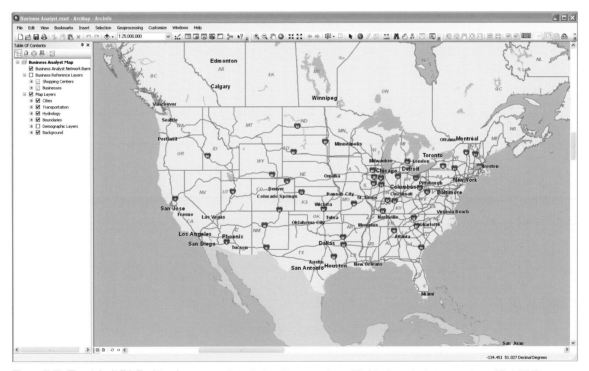

Figure 7-7. The default BA Desktop base map. Data displayed in screenshots of Esri Business Analyst are courtesy of Esri; US Census Bureau; Infogroup; Bureau of Labor Statistics; Applied Geographic Solutions, Inc.; Directory of Major Malls, Inc.; GfK MRI; and Market Planning Solutions, Inc.

In this example, the project team will geocode a spreadsheet containing the address locations of community colleges throughout California (almost any location data could be substituted in its place). Demonstrating these procedures illustrates the straightforward manner in which geocoding tasks can be handled (your interface may vary, depending on the versions of BA Desktop and ArcGIS you are using). Figure 7-7 shows how Business Analyst appears upon launch.

Figure 7-8 shows a portion of a spreadsheet of community colleges that was added to Business Analyst's Table of Contents (the vertical listing of map layers is shown on the left side of figure 7-7).

"Geocode Addresses" can be selected by right-clicking the Excel file in the Table of Contents to view the resulting context-sensitive menu shown in figure 7-9.

Next, the "Choose an Address Locator to use" dialog box appears. BA Desktop contains a built-in address locator tool titled "USA Geocoding Service."

Next, the appropriate attribute fields from the input spreadsheet are selected, using the drop-down lists, as shown in figure 7-11.

FID	Shape	College	Address	City	State	Zip	Phone	Fax
0	Polygon	Ohlone College	43600 Mission Blvd.	Fremont	CA	94539	510-659-6000	none
1	Polygon	Gavilan College	5055 Santa Teresa Blvd	Gilroy	CA	95020	408-848-4800	408-848-4801
2	Polygon	Cuyamaca College	900 Rancho San Diego Pkw	El Cajon	CA	92019	619-660-4000	619-660-4399
3	Polygon	Imperial Valley College	380 East Aten Road	Imperial	CA	92251	760-352-8320	760-355-2663
4	Polygon	MiraCosta College	One Barnard Dr.	Oceanside	CA	92056	760-757-2121	760-795-6609
5	Polygon	Palomar College	1140 W. Mission Rd.	San Marcos	CA	92069	760-744-1150	760-744-8123
6	Polygon	San Diego City College	1313 - 12th Ave.	San Diego	CA	92101	619-388-3400	619-388-3501
7	Polygon	San Diego Miramar College	10440 Black Mountain Rd.	San Diego	CA	92126	619-388-7800	619-388-7915
8	Polygon	San Diego Mesa College	7250 Mesa College Dr.	San Diego	CA	92111	619-388-2600	619-388-2929
9	Polygon	Southwestern College	900 Otay Lakes Rd.	Chula Vista	CA	91910	619-421-6700	619-482-6413
10	Polygon	Butte College	3536 Butte Campus Dr.	Oroville	CA	95965	530-895-2511	530-895-2345
11	Polygon	Feather River College	570 Golden Eagle Avenue	Quincy	CA	95971	530-283-0202	530-283-3757
12	Polygon	Lassen College	PO Box 3000	Susanville	CA	96130	530-257-6181	530-251-8872
13	Polygon	Mendocino College	1000 Hensley Creek Rd.	Ukiah	CA	95482	707-468-3000	707-468-3120
14	Polygon	College of the Redwoods	7351 Tompkins Hill Rd.	Eureka	CA	95501	707-476-4100	707-476-4400
15	Polygon	Shasta College	PO Box 496006	Redding	CA	96049	530-242-7500	530-225-4990
16	Polygon	College of the Siskiyous	800 College Ave.	Weed	CA	96094	530-938-5555	530-938-5227
17	Polygon	Lake Tahoe Community College	One College Dr.	South Lake Tahoe	CA	96150	530 541-4660	530 541-7852
18	Polygon	American River College	4700 College Oak Dr.	Sacramento	CA	95841	916-484-8011	916-484-8674
19	Polygon	Cosumnes River College	8401 Center Pkwy	Sacramento	CA	95823	916-691-7451	916-691-7330
20	Polygon	Folsom Lake College	10 College Parkway	Folsom	CA	95630	916-608-6993	916-608-6584
21	Polygon	Sacramento City College	3835 Freeport Blvd.	Sacramento	CA	95822	916-558-2111	916-650-2909
22	Polygon	Napa Valley College	2277 Napa-Vallejo Hwy.	Napa	CA	94558	707-253-3000	707-253-3015
23	Polygon	Santa Rosa Junior College	1501 Mendocino Ave.	Santa Rosa	CA	95401	707-527-4011	707-527-4816
24	Polygon	Sierra College	5000 Rocklin Rd.	Rocklin	CA	95677	916-781-0540	916-789-2855
25	Polygon	Solano Community College	4000 Suisun Valley Rd	Fairfield	CA	94534	707-864-7000	707-864-7213
26	Polygon	Yuba College	2088 North Beale Road	Marysville	CA	95901	530-741-6700	530-741-3541
27	Polygon	Woodland Community College	2300 E. Gibson Road	Woodland	CA	95776	530-661-5700	
28	Polygon	Diablo Valley College	321 Golf Club Rd.	Pleasant Hill	CA	94523	925-685-1230	925-685-1551
29	Polygon	Los Medanos College	2700 E. Leland Rd	Pittsburg	CA	94565	925-439-2181	925-427-1599
30	Polygon	Contra Costa College	2600 Mission Bell Dr.	San Pablo	CA	94806	510-235-7800	510-236-6768
31	Polygon	College of Marin	835 College Ave.	Kentfield	CA	94904	415-457-8811	415-485-0135
32	Polygon	College of Marin Indian Valley Campus	1800 Ignacio Blvd	Novato	CA	94947	415-457-8811	
33	Polygon	Berkeley City College	2050 Center St.	Berkeley	CA	94704	510-981-2800	510-841-7333
34	Polygon	Laney College	900 Fallon St	Oakland	CA	94607	510 834 5740	510 464 3528

(0 out of 112 Selected)

CampusesWithDistrictInfo

Figure 7-8. BA Desktop will be used to geocode the spreadsheet of the community colleges. Note the attributes, including college name, address, city, state, and ZIP Code.

Figure 7-9. The geocoding process in BA Desktop begins by right-clicking the community college spreadsheet listing in the Table of Contents. Data displayed in screenshots of Esri Business Analyst are courtesy of Esri; US Census Bureau; Infogroup; Bureau of Labor Statistics; Applied Geographic Solutions, Inc.; Directory of Major Malls, Inc.; GfK MRI; and Market Planning Solutions, Inc.

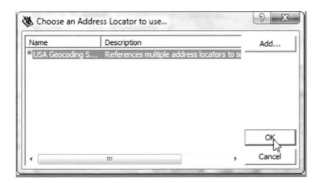

Figure 7-10. Selecting an address locator in BA Desktop. The software includes a locator for the entire United States, making geocoding anywhere in the country easy. Esri provides built-in locators that cover other countries as well.

Figure 7-11. The geocoding dialog box in BA Desktop; note the four address component fields listed. In this case, the attribute field names in the community college spreadsheet match the field names sought by the dialog box.

The geocoding process proceeds with a dialog box that displays the progress of the operation and is completed when the message appears, as shown in figure 7-12.

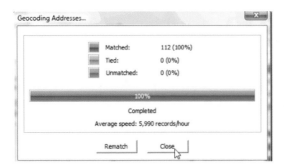

Figure 7-12. The results show that all 112 of the community colleges were successfully geocoded. If addresses fail to match, the interactive rematch process can be commenced by clicking the "Rematch" button. This allows the GIS user to improve match possibilities by, for example, correcting spelling mistakes or incorrect addresses present in the input spreadsheet.

The results of geocoding are displayed in the map, and the Table of Contents will now include a new layer of the geocoded points, as shown in figure 7-13. This new layer can be saved for use in other projects.

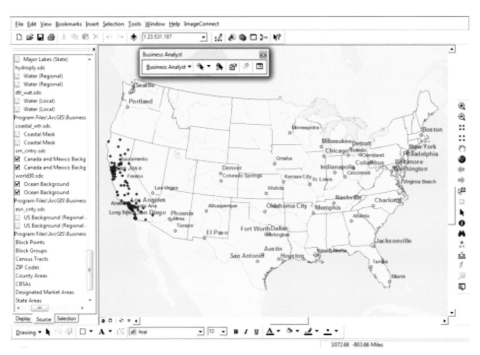

Figure 7-13. The results of the geocoding operation in BA Desktop result in the map display showing the 112 California community colleges. Data displayed in screenshots of Esri Business Analyst are courtesy of Esri; US Census Bureau; Infogroup; Bureau of Labor Statistics; Applied Geographic Solutions, Inc.; Directory of Major Malls, Inc.; GfK MRI; and Market Planning Solutions, Inc.

Opening the attribute table for this new map layer (figure 7-14) reveals that x,y coordinates, representing the latitude and longitude values for the geocoded points, have been appended (they were not part of the original spreadsheet). This is a "value-added" outcome of the BA Desktop geocoding process.

FID	Shape	College	Latitude	Longitude
0	Point	Mt. San Jacinto College	33.805985	-116.966461
1	Point	Barstow College	34.871766	-117.025582
2	Point	Mt. San Antonio College	34.044712	-117.841854
3	Point	Cerro Coso Community College	35.567557	-117.670305
4	Point	Bakersfield College	35.408219	-118.971756
5	Point	Victor Valley College	34.475942	-117.260085
6	Point	College of the Siskiyous	41.411168	-122.390628
7	Point	Merced College	37.335747	-120.473628
8	Point	College of the Redwoods	40.697729	-124.196584
9	Point	American River College	38.649536	-121.346744
10	Point	Lake Tahoe Community College	38.926866	-119.971606
11	Point	Ohlone College	37.528501	-121.91479
12	Point	Santa Barbara City College	34.405718	-119.698644
13	Point	Santa Rosa Junior College	38.455357	-122.721238
14	Point	Palomar College	33.150356	-117.181754
15	Point	West Hills College Coalinga	36.149251	-120.356705
16	Point	Butte College	39.64912	-121.646105
17	Point	Cabrillo College	36.989431	-121.924658
18	Point	Yuba College	39.12439	-121.537885
19	Point	Woodland Community College	38.658892	-121.735365
20	Point	Long Beach City College	33.830464	-118.136213
21	Point	Cosumnes River College	38.454032	-121.422708
22	Point	Folsom Lake College	38.662641	-121.126633
23	Point	Sacramento City College	38.540838	-121.488414
24	Point	Feather River College	39.950021	-120.969425
25	Point	College of Marin	37.953228	-122.549126
26	Point	College of Marin Indian Valley Campus	38.075356	-122.579651
27	Point	Fresno City College	36.767517	-119.793695
28	Point	Porterville College	36.048336	-119.012529
29	Point	Reedley College	36.60652	-119.461009
30	Point	Sierra College	38.793344	-121.211107
31	Point	Monterey Peninsula College	36.590258	-121.885344
32	Point	Allan Hancock College	34.942201	-120.421795
33	Point	Antelope Valley College	34.676973	-118.187373
34	Point	San Bernardino Valley College	34.08751	-117.310738
35	Point	Crafton Hills College	34.039298	-117.100678

(0 out of 112 Selected)

Figure 7-14. Attribute table of the successfully geocoded community colleges map layer. Note that x,y (latitude and longitude) values have been automatically appended to the table during the geocoding process.

Geocoding is possible even without complete addresses and can be employed with other information, including latitude/longitude coordinates obtained from a GPS device (as seen in chapter 6), a listing of customer ZIP Codes, property identification numbers, city names, and other data.

The previous examples demonstrate the "ease of use" nature of BAO and the "full control" nature of BA Desktop. An additional advantage of BA Desktop is its tight integration with ArcGIS, allowing users to leverage the powerful geoprocessing tools available there. More specifically, while BAO is easy to learn and use, two of its limitations at the time of this writing are that it can only work with polygon data that is contiguous (such as census tracts), and it can only geocode a maximum of one hundred records at one time. On occasion, economic development practitioners will find the need to overcome these limitations; for example, an analyst may be interested in a number of opportunity areas in multiple cities that are clearly not contiguous. Or there might be a need to geocode a very large database of thousands of addresses without worrying about input limits. In such instances, BA Desktop provides a number of tools to work with large datasets and the ability to precisely customize the geocoding procedures.

Incorporating ModelBuilder into geocoding

Chapter 3 advocated for the use of ModelBuilder to preserve geoprocessing workflows and facilitate coordination within the project team. Since geocoding is a commonly used geoprocessing application, geocoding can be represented in ModelBuilder. In this example, a project team wants to geocode a spreadsheet of San Jose hotel addresses, shown in figure 7-15, for a tourism study (again, your interface may vary from the one shown here, depending on the version of ArcGIS you are using).

Address	City	State	ZipCode
1471 North 4th Street	San Jose	CA	95112
1560 North First St	San Jose	CA	95112
1602 Crane Court	San Jose	CA	95112
233 Santa Clara	San Jose	CA	95128
2660 Monterey Rd	San Jose	CA	95111
282 Almaden Avenue	San Jose	CA	95113
2885 Seaboard Avenue	San Jose	CA	95131
455 South 2nd Street	San Jose	CA	95113
55 Old Tully Road	San Jose	CA	95111
6199 San Ignacio Ave	San Jose	CA	95119

Figure 7-15. A spreadsheet of addresses to be geocoded: a partial list of hotels in San Jose, California.

Figure 7-16 shows how the process of geocoding an address list of hotels in San Jose might be represented in ModelBuilder. Note the two tools (yellow) reflecting two geoprocesses. The first, Create Address Locator, is used to create an address locator (a user-created file that stores the address data and the map features to which the addresses will be located; San Jose roads, in this case). The second, Geocode Addresses, is used later in the model to execute the geocoding operation. Also note the input data (San Jose roads and the hotel address list), represented by blue ovals, as well as the output datasets (address locator file and, in the end, a geocoded hotels dataset), represented by green ovals.

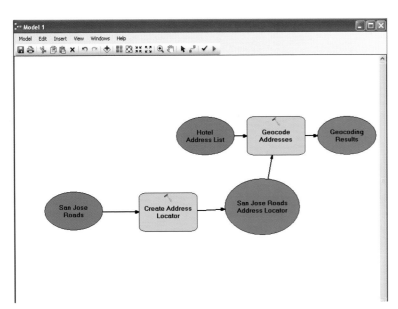

Figure 7-16. A geocoding process shown in ModelBuilder; the blue inputs represent a spreadsheet of hotel addresses and a reference dataset of San Jose roads, against which the addresses will be geocoded using an ArcGIS address locator. The end result is a new dataset representing the geocoded hotel locations.

If another list needs to be geocoded later, ModelBuilder can be used to swap out the existing hotel model element and replace it with a new support services address dataset, then rerun the model to generate a new set of geocoded points. A new address locator can be created and customized for a community, or the countrywide address locator from Esri can be used to create a generic geocoding model for geocoding any address in the country. This is a useful tool for projects scattered throughout the nation.

Geocoding without specific addresses

The previous example demonstrated the process for geocoding hotels with complete address information, including the street number, street name, city, and state. Addresses with this level of specificity allow for precise geocoding results, assuming, of course, that the addresses are properly spelled and numbered in the input dataset. However, locations may need to be geocoded using a dataset that contains a lower level of locational specificity. Sometimes data providers deliberately remove portions of addresses from databases to protect anonymity. Some datasets commonly used by economic development analysts, such as the Economic Census of 2002 and 2007, will identify industries only down to the 5-digit ZIP Code level. Also, some trade associations provide data only down to the 3-digit ZIP Code level. Likewise, sometimes the US Census Bureau withholds data—such as owner's assessment of home value—if there are too few owner-occupiers in a census block group, as is the case in 75 of the 575 census block groups in San Francisco.

Geocoding can be performed with these limitations, understanding that the end result may not always be as geographically specific as one that resulted from specific addresses. The next example considers a case where the location information is presented as pairs of latitude/longitude coordinates. The example in figure 7-17 uses a different version of the San Jose hotels spreadsheet from the one presented earlier (your interface may vary, depending on the versions of software you are using).

Hotel	City	State	ZipCode	Latitude	Longitude
Ramada	San Jose	CA	95113	37.329673	-121.884516
Radisson San Jose Airport	San Jose	CA	95112	37.364304	-121.907436
Hampton Inn Suites	San Jose	CA	95111	37.301439	-121.857517
Hotel DeAnza	San Jose	CA	95126	37.338656	-121.885744
Holiday Inn Express	San Jose	CA	95111	37.295302	-121.852759
Crowne Plaza Hotel	San Jose	CA	95113	37.32991	-121.891441
La Quinta Inn	San Jose	CA	95131	37.37947	-121.939379
Staybridge Suites	San Jose	CA	95112	37.372971	-121.911674
Extended Stay America	San Jose	CA	95119	37.235125	-121.77666
Homestead Studio Suites	San Jose	CA	95112	37.364359	-121.911348

Figure 7-17. Hotel data, including latitude and longitude coordinates.

Coordinate pairs might be derived, for example, from GPS units after a technician completes field work and downloads the unit's data to a spreadsheet file, as seen in chapter 6. ArcGIS provides a tool called "Add XY Data" for geocoding such data; figure 7-18 shows the location of this option, and figure 7-19 shows the input dialog box (your interface may vary slightly, depending on the version of ArcGIS software used). The corresponding dialog box lets users choose the attribute field names present in the input dataset that correspond to the "X field" (longitude) and the "Y field" (latitude).

When the Add XY Data process is complete, a new geocoded map layer will be added to ArcGIS representing, in this example, the San Jose hotel locations.

Figures 7-18 and 7-19. The location of the Add XY Data function in ArcGIS and the corresponding dialog box. This function facilitates geocoding in the absence of specific addresses when only coordinates (such as latitude/longitude pairs) of locations are available.

Further analysis using geocoded points

Geocoding locations is clearly an important task for numerous economic development analyses, but more can be done in the way of spatial analysis once the location points are present in a project map. BAO was used earlier in this chapter, for example, to generate drive-time polygons from SJSU's geocoded location. Such analysis is useful for answering geographic questions such as "How many consumers live within a five-minute drive of this location?" or "Where might a new coffeehouse be located within a ten-minute drive of a neighborhood with certain demographic characteristics?"

But what if the interest was in a more fine-grained analysis that didn't rely on drive times but on walking distances, or distances that could be traversed by bicycle or public transit? A project team has the choice of BAO, with its ease of use, or the full control offered within ArcGIS. Whereas BAO automates the creation of drive-time areas "behind the scenes," ArcGIS users have access to the same technology through the ArcGIS Network Analyst extension that extends the core functionality of ArcGIS to facilitate more customization for project needs. Network Analyst facilitates a number of analysis types, including calculations of the shortest path between two locations based on distance or time limits, and the generation of "service areas" (synonymous with drive-time areas, but more customizable to account for other forms of transportation besides cars). Network Analyst can also be used to find the closest facility to a given location based on the actual road network; in economic development projects, this is useful for finding, say, the locations of businesses closest to customer or competitor locations. Figure 7-20 shows the portions of northeastern San Francisco that are within a five-minute walk of geocoded coffeehouses.

Building upon the discussion of mixed-use, transit-oriented developments (TOD) begun in chapter 6, the next scenario focuses on an economic development planner for a city with a high proportion of renters, who rely on modes of transport other than cars to a greater degree than do their homeowner counterparts in the suburbs. The city wants to promote TODs as a way to reduce commuting costs by locating a mix of land uses in places that capitalize upon investments in bicycle lanes, local bus service, and light rail access. For that reason, the planner has been tasked with finding suitable sites for a new downtown TOD project. More specifically, the planner wants to identify parts of the city that can be easily accessed by different modes of transport. Specifically, the goal is to identify areas that can be accessed from possible TOD sites within ten minutes using the various transportation options, since pedestrians, cyclists, motorists, and transit passengers all travel at varying speeds. In this scenario, BAO's easy to use drive-time functionality would be too limiting since it assumes that all travel occurs by car. However, Network Analyst in ArcGIS can be employed to customize the analysis for various transportation modes.

An example of resulting service areas might look as shown in figure 7-21; note the wide variation in service areas based on transport mode.

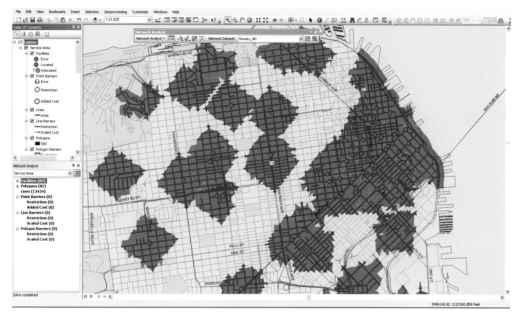

Figure 7-20. Portions of San Francisco that are within a five-minute walk (generally, about one-quarter mile) of geocoded coffeehouses (red dots), using ArcGIS Network Analyst. The walkable areas are shown in light brown, and the corresponding streets are shown in dark brown. Note the ArcGIS Network Analyst toolbar toward the top of the display, where all network functions can be customized for unique project needs, such as specifying walking, driving, transit, or bicycling travel speeds.
Courtesy of City and County of San Francisco Enterprise GIS

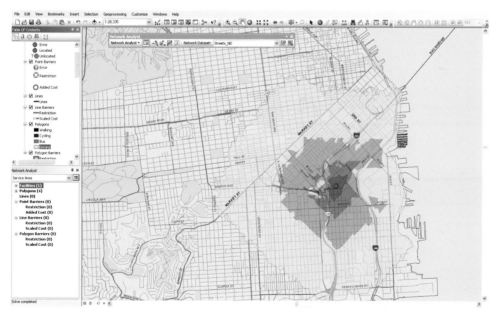

Figure 7-21. Portions of San Francisco that are within a ten-minute distance from a proposed, geocoded TOD site (red dot). The polygons emanating from the site reflect, in order from smallest to largest (and from darkest to lightest purple), areas that could be accessed during that time by walking, bicycling, bus, and car, respectively. ArcGIS Network Analyst facilitates such fine-grained network analysis within the ArcGIS geoprocessing environment. Courtesy of City and County of San Francisco Enterprise GIS

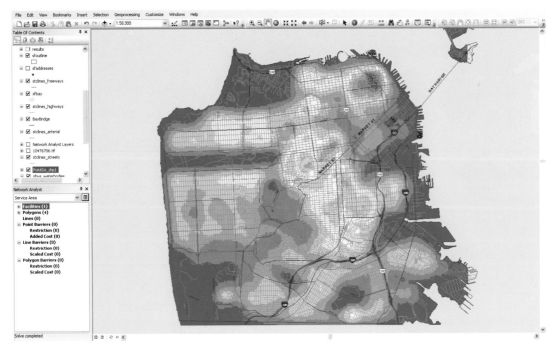

Figure 7-22. Clustering analysis using geocoded points of residence locations in San Francisco. Bluer shades reflect lower densities of residences, and redder shades represent higher densities. ArcGIS is equipped with geoprocessing tools to make the creation of such maps straightforward while opening the door to more advanced spatial statistical analysis, explored in chapter 8.
Courtesy of City and County of San Francisco Enterprise GIS.

Other interesting uses of ArcGIS tools to interpret geocoded points include the ability to produce maps that illustrate concentrations, or clusters, of these points. Consider figure 7-22, in which geocoded residence locations in San Francisco are reflected as clusters. Areas trending from blue to red represent increasingly denser concentrations of residences. Such an analysis might be useful for an economic development study dedicated to finding the most suitable site for a new TOD that would attract those wishing to live in a dense, walkable neighborhood.

The availability of geocoded data allows the use of many of the geoprocessing tools in ArcToolbox, including many of the spatial statistics and regression tools examined in chapter 8.

Summary

The ability to exploit address or other locational information is essential in carrying out economic development projects. The possibilities for geospatial analysis inherent in any list of locations are revealed once the geocoding process produces a digital pin map. This chapter demonstrated the process of geocoding using BAO, BA Desktop, and ArcGIS and compared the relative "ease of use" and "full control" offered by these tools to conduct geocoding. Geocoding operations can be built into geoprocessing models that can be easily updated. The chapter briefly explored advanced uses of geocoded points for transportation cost analysis using ArcGIS Network Analyst and provided an example of cluster maps that can be generated with the geocoded points serving as input.

8

Statistics and statistical methods in economic development using GIS

Objectives

- To show why statistical analyses of GIS data are important to economic development analysis
- To show how some spatial statistics tools of ArcGIS software can be used in economic development analyses, including the following:
  - Hot spot analysis
  - Mean center
  - Median center
  - Standard distance
  - Standard deviational ellipse
  - (Global) spatial autocorrelation
  - Regression analysis

Clustering: Point patterns versus hot spot analysis

The last chapter ended with a map showing residence densities in San Francisco and indicating a concentration, or clustering, of coffeehouses, with a density pattern of points. The perception of clustering in a density map can be affected by many factors—especially the neighborhood size and the number and types of categories into which the phenomenon is divided. For example, the maps in figures 8-1a to 8-1d give different senses of the concentration of the phenomenon (concentration of biotechnology firms) with high concentrations indicated in red and low concentrations indicated in blue. However, this is only a matter of perception—the data for these maps are exactly the same. The first two maps show the data displayed at the county level, divided into five and three categories, respectively. The next two maps show the data displayed at the census tract level, divided again into five and three categories, respectively.

Figures 8-1a to 8-1d. The same data on the distribution of biotechnology firms in the San Francisco Bay Area in the style of a hot spot map (with red being high and blue being low), shown for different levels of aggregation (figures 8-1a and 8-1b for county level, and figures 8-1c and 8-1d for census tract) and different numbers of categories of classification (category three for figures 8-1b and 8-1d and category five for figures 8-1a and 8-1c). Courtesy of US Census and BayBio.

Such density patterns are suggestive but "visual only." This means that the pattern is in the eye of the beholder. In the words of leading researchers in the area, "Humans are adroit at recognizing patterns, even when no patterns exist."[1]

The methods of spatial statistics allow us to examine a pattern in terms of statistical significance. These methods allow us to make an *objective assessment* about clustering rather than a *subjective perception* of a pattern.

There are two related issues concerning clustering from a statistical point of view. First, for the overall pattern in a study area, does clustering occur? This is a global question concerning the entire study area. A tool of spatial statistics, the global Moran's I (named for its developer, the statistician Patrick A. P. Moran), allows us to say, for a given pattern, whether there is statistically significant clustering. Clustering is statistically significant if there is a very small probability that the pattern arose by chance. The answer to the global question is either "yes" or "no."

While it is important to know whether there is statistically significant clustering, the global test does not indicate where the clusters are. Another tool of spatial statistics, hot spot analysis, is used to determine where the clusters are located based on statistical significance. Hot spot analysis gives an objective basis for determining where clusters are. The global Moran's I test is applied, and a hot spot analysis is undertaken later in this chapter.

Statistics and GIS

The variety of statistical tools in ArcGIS generally falls into some familiar categories in statistics. Some tools describe central tendency—the mean and median of a distribution. Some describe dispersion—how a variable is distributed. Others describe the relationships among variables, such as correlation and regression. The spatial statistical tools in ArcGIS address central tendency, dispersion, and relationships among variables but from a spatial point of view. ArcGIS also has some traditional statistical tools that will be mentioned in this chapter.

Basic statistical concepts

This sidebar introduces three basic statistical concepts: the distribution of a variable, measures of central tendency, and measures of dispersion.

Distribution of a variable

The variation in values of a variable or field—for example, the variation in median household income across census tracts—can be described by a histogram or frequency distribution (see the sidebar, "Statistics of a Field in ArcGIS," in this chapter). The distribution may be symmetric (as in the case of the "bell curve" of the normal distribution), or it may be skewed (as is the distribution of income).

Central tendency—mean and median

There are two main measures of central tendency (or two different senses of the term "average"). The most common measure of average is the arithmetic mean: the sum of the values of all the observations divided by the number of observations. When the distribution of the variable is skewed, it may be deceptive to rely on the mean. The distribution of median household income by census tract shown in the in this chapter on "Statistics of a Field" is skewed to the right—there are a few census tracts with very high levels of income. These income levels are so high that they disproportionately affect the arithmetic mean—they "pull the mean up." This is typical of income statistics, whether of census tracts or individuals. A few wealthy individuals pull up the mean income. Likewise, statistics for housing prices and duration of unemployment are similarly skewed to the right—a few observations with very large values. Relatively few "monster" houses and relatively few individuals who are unemployed for exceptionally long periods pull up the respective mean values. In these cases, the median value may give a truer sense of average. For median income, 50 percent of the observations have a greater value, and 50 percent of the observations have a lower value. The median is unaffected by extreme values. While the mean value of household incomes in the census tracts shown in the sidebar, "Statistics of a Field in ArcGIS," is $80,962, the median value is only $75,272.

Dispersion—variance and standard deviation

While a frequency distribution (as illustrated in figure 8S-1 generated by the Field Statistics functionality in ArcGIS) is very useful for understanding how a variable is distributed, it is often convenient to have a single statistic that summarizes the dispersion of a variable. One such statistic is the standard deviation. The standard deviation has a convenient interpretation in the case of a variable that is distributed approximately normally. The range of values encompassed by the mean plus and minus one standard deviation includes about 67 percent of the values in the distribution. The range of value encompassed by the mean plus and minus two standard deviations includes about 95 percent of the values in the distribution. The range of values encompassed by the mean plus and minus three standard deviations includes about 99 percent of the values in the distribution.

Statistics of a field in ArcGIS

It is easy to determine descriptive statistics of any field (variable) that has been read into ArcGIS. By opening the attribute table of a layer in a map document, the fields and records behind the map display are revealed (your interface may vary, depending on the version of software used). The Statistics of a Field function can easily be invoked to show both the central tendency (mean) and dispersion (as shown in a graph of the distribution) of any variable.

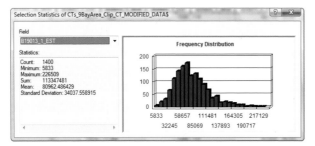

Figure 8S-1. The statistics of median household income in the San Francisco Bay Area by census tracts. Courtesy of US Census.

Several statistical issues arise within a spatial context, which do not arise in traditional nonspatial statistics. These spatial statistical issues will be contrasted with the more familiar traditional statistical measures when possible.

Measures of central tendency

The most familiar measures of central tendency are mean and median. The mean is the arithmetic average—for finding the average on an exam the scores are added, then divided by the number of people who took the exam. Technically, this is the unweighted (or equally weighted) average. Everyone's score counts the same. Those with better scores on the previous exam do not count more or less than others. Sometimes, the interest is in a weighted average. For example, the course average may depend on scores on three exams, with the first exam counting 20 percent, the second exam counting 30 percent, and the third exam counting 50 percent.

In a spatial context, the unweighted or weighted average *location* also can be determined. The average can be either the mean or the median, either weighted or unweighted. As this chapter will explain, the mean and median locations have special interpretations in spatial statistics.

In the ArcGIS Spatial Statistics toolbox, there is a **Mean Center tool,** which can be used to determine either the unweighted or weighted mean center—the equivalent of the arithmetic average location. Any variable of interest can be selected for determining the weighting. Figure 8-2 shows three mean centers of the 341 census tracts in Santa Clara County—the unweighted mean center, the mean center weighted by population, and the mean center weighted by income. These are three different points. The unweighted mean center is the arithmetic average point from the **centroid** (center of gravity) of each census tract. Centroids are used to determine distances from polygon shapes to points, although within ArcGIS distances can be determined to the closest point in a polygon. (This can be done using the Near tool. This tool is used later in this chapter to determine distances from biotechnology firms to research universities.)

The distribution of population and the distribution of median income are not identical across census tracts. Weighting the mean center by median household income will "pull" the central point in the direction of greater income. Weighting the mean center by population will "pull" the central point in the direction of greater population.

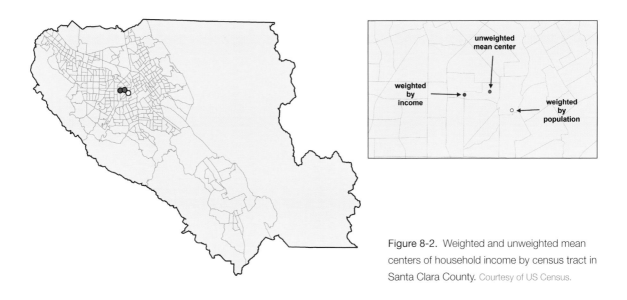

Figure 8-2. Weighted and unweighted mean centers of household income by census tract in Santa Clara County. Courtesy of US Census.

The choropleth maps in figures 8-3a and 8-3b, which describe the distribution of median household income and of population by census tract in Santa Clara County, also show that the "pull" in the respective cases appears to be in the correct direction—the direction suggested by choropleth maps showing the distribution of income and population, respectively.

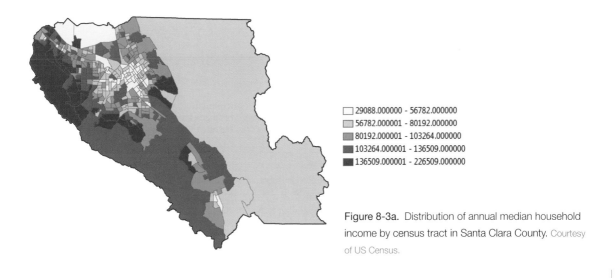

☐ 29088.000000 - 56782.000000
☐ 56782.000001 - 80192.000000
☐ 80192.000001 - 103264.000000
☐ 103264.000001 - 136509.000000
■ 136509.000001 - 226509.000000

Figure 8-3a. Distribution of annual median household income by census tract in Santa Clara County. Courtesy of US Census.

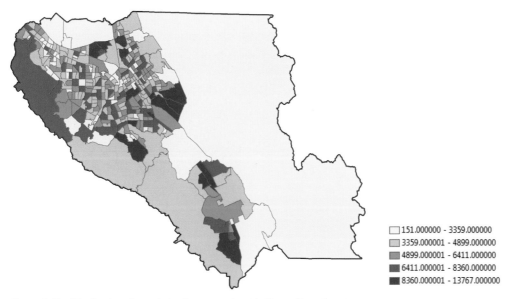

	151.000000 - 3359.000000
	3359.000001 - 4899.000000
	4899.000001 - 6411.000000
	6411.000001 - 8360.000000
	8360.000001 - 13767.000000

Figure 8-3b. Distribution of population by census tract in Santa Clara County. Courtesy of US Census.

Another example will illustrate the concept of the **central feature** rather than the mean center. Figure 8-4 shows California community colleges.[2] This is *point* vector data, in contrast to the *polygon* vector data used in the Santa Clara County census tracts example. The mean center has been identified (weighted by full-time equivalent enrollment in credit courses). This enrollment-weighted mean center is at a point that does not correspond to any campus location. Perhaps the interest is in *the campus,* which is the enrollment-weighted mean. That would be the enrollment-weighted central feature (determined using the Central Feature tool in the Spatial Statistics toolbox). Figure 8.4 shows a map of California community colleges and the enrollment-weighted mean center and enrollment-weighted central feature.

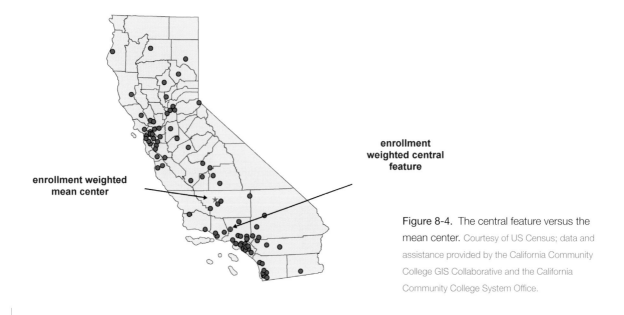

enrollment weighted central feature

enrollment weighted mean center

Figure 8-4. The central feature versus the mean center. Courtesy of US Census; data and assistance provided by the California Community College GIS Collaborative and the California Community College System Office.

Sometimes the median is more significant as a measure of central tendency, especially if the distribution is skewed. In a spatial context, there is another reason to be interested in the median. In urban economics, the principle of median location[3] is often used to explain the location of a retail establishment. The median location minimizes the aggregate distance that customers travel to reach a retail outlet.

In the community college example, the unweighted median center and the enrollment-weighted median center will be determined using the **Median Center tool**. Each has a distinct spatial interpretation. The unweighted (or equally weighted) median center is the point that minimizes the aggregate distance traveled, assuming that each campus is weighted equally. This is the place to hold a meeting if each campus sends one representative. On the other hand, if each campus sends representatives proportional to enrollment, the task would be to find the point that minimizes the aggregate travel of all representatives. This is the enrollment-weighted median center. Figure 8.5 shows these median centers.

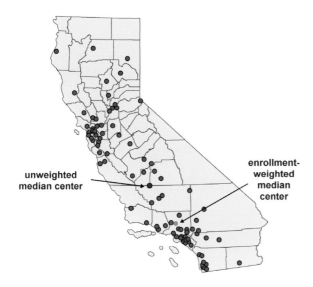

Figure 8-5. The enrollment-weighted center versus the unweighted median center. Courtesy of US Census; data and assistance provided by the California Community College GIS Collaborative and the California Community College System Office.

Measures of dispersion and distribution

The distribution of a variable shows how close or how far the values vary from the center and whether they are more or less evenly distributed around that center or if they are biased (skewed) one way or another. Typically, the dispersion of a variable about the mean is described in terms of the standard deviation. Several tools in ArcGIS can be used to describe the spatial distribution of a variable.

When assessing the dispersion of points, the **Standard Distance tool** is useful. This gives a radius that encompasses one, two, or three standard deviations around the unweighted or weighted mean center. These standard deviational circles encompass, respectively, roughly 67 percent, 95 percent, and 99 percent of the points. The size of these circles offers some sense of how narrowly or widely dispersed the points are. Figure 8.6 shows the one, two, and three standard deviation standard distances from the median income-weighted mean center of Santa Clara County census tracts.

Chapter 1 used the Standard Deviational Ellipse tool on point data to show the directionality of the distribution of biotechnology firms in the San Francisco Bay Area. The same can be done for the distribution of median household income in Santa Clara County. The center of each ellipse will be the mean center of the distribution, but instead of forcing the determination of standard deviations to fall into a circle, now they can fall into an ellipse (with a circle as an unlikely special case).

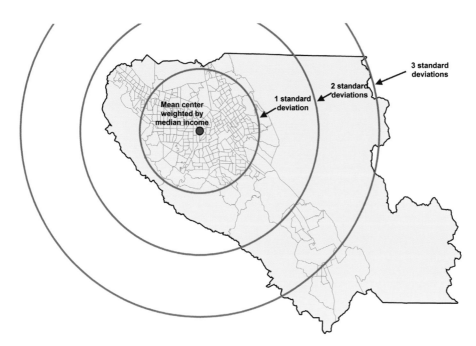

Figure 8-6. The one, two, and three deviational standard distances from the mean center weighted by income (blue dot).
Courtesy of US Census.

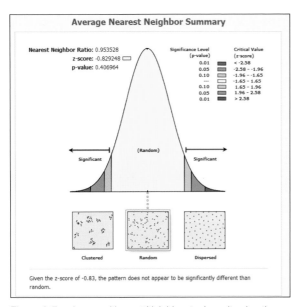

Figure 8-7a. Average Nearest Neighbor tool result using the smaller default area.

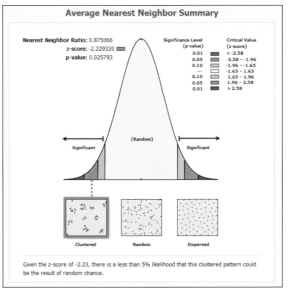

Figure 8-7b. Average Nearest Neighbor tool result using the entire state as the area.

Another spatial statistics tool, the **Average Nearest Neighbor,** computes an index that compares the distribution of points with a hypothetical random distribution. The Average Nearest Neighbor does not provide for a weight. It just examines the distribution of points across an area and determines whether a given distribution of points is statistically significantly different from a random distribution. The area in which the points are located is critical to the determination. The same set of points might be randomly distributed or clustered, depending on the area. For example, the seven public community colleges located in Santa Clara County are not clustered there. (Remember that the Average Nearest Neighbor index does not account for such factors as size of population. Indeed, the same community colleges are clustered around more heavily populated areas of Santa Clara County, as seen in figure 1-10 in chapter 1.) However, the same seven community colleges *are* clustered when the entire state of California is considered. With respect to this larger area, the location of the seven points is statistically improbable as a random distribution. Figures 8-7a and 8-7b show the output of the Average Nearest Neighbor tool.

Modeling spatial relationships among variables

Economic development officials rely on relationships among variables to formulate, implement, and assess policies. This section examines some of the main ways in which relationships among variables are discussed and presented in statistical analysis applied to economic development. The focus is on measuring and modeling spatial relationships among variables.

Correlation

Correlation is a measure of the *linear* association between two variables. Correlation between two variables ranges from zero (no linear association) to one (perfect linear association). It is possible to have a perfect *non-linear* association, which results in less than one computed correlation. Consider, for example, two variables, X and Y, related by the equation $Y=X^2$. The two variables are perfectly correlated, but not perfectly correlated *linearly*. For values of X from 1 to 10 and the corresponding values of Y, the correlation coefficient would be 0.9745586. On the other hand, if the two variables had been related by the equation $Y=a+bX$, then the correlation coefficient would be 1 if $b>0$ and -1 if $b<0$. In a spatial context, the correlation between median income in a census tract and percent owner-occupied housing in a census tract for the 341 census tracts of Santa Clara County California is 0.63. Figure 8-8 shows the **scatter diagram (scatter plot)** of median income and percent owner-occupied. Correlation is *suggestive* of a relationship between the variables, but there are numerous examples where the suggestion is mistaken. This high degree of correlation is not surprising given that qualification standards for a mortgage (housing purchase) loan are tied to income.

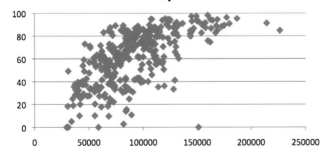

Figure 8-8. Scatter diagram showing median income and percent owner-occupied by census tract for Santa Clara County, California. Courtesy of US Census.

Autocorrelation

Autocorrelation is correlation of a variable with itself. A variable can be correlated with itself either through time (temporal autocorrelation) or across a geographic area (spatial autocorrelation). Either kind of autocorrelation can be somewhat problematical in terms of classical regression analysis. At the same time, spatial autocorrelation, which is the primary concern here, is inherent in geographical analysis. As the statistician and geographer Daniel Griffith observed, "It is axiomatic to geography that events in space are not random. When geographers examine the spatial distribution of phenomena, they attempt to understand and explain patterns as resulting from processes, causal linkages between events in time and space."[4] This section will discuss tools in ArcGIS for measuring spatial autocorrelation.

The first example considers temporal autocorrelation—correlation of a variable with itself through time. What is a good estimate of tomorrow's stock price? Today's stock price! Stock prices change daily, but often they do not change much from day to day. A high degree of correlation is expected between today's stock price and tomorrow's stock price. This is autocorrelation—the same variable, the stock price, is correlated with itself temporally—displaced by one day. Actually, stock index prices displaced one day over a period of a year have a correlation of about 0.9. The longer the displacement—say, one week instead of one day—the lower the correlation. Computing the correlation of stock prices today with stock prices a week ago, for a year, would give a correlation of about 0.7. The correlation would fall to about 0.4 for a greater temporal displacement of, say, two weeks.

For many variables, the temporal autocorrelation of a variable is greater, the closer in time they are observed. There is an important caveat: the time series may exhibit *periodicity*—there may be natural cycles in the data. So the correlation may grow stronger, and the temporal displacement greater, when the displacement coincides with the period of the natural cycle. *Time series analysis* deals with these deep issues.

With spatial autocorrelation, the measure of displacement is spatial rather than temporal. Spatial autocorrelation is more complicated than temporal autocorrelation because the spatial relationship is multidimensional and can be measured in several ways. For example, distance can be measured by the Euclidean metric, the Manhattan distance, or contiguity.

Figure 8-3a showed median income by census tracts in Santa Clara County. It appears that median income values are spatially correlated—high median income census tracts are close to other high median income census tracts and low median income census tracts are close to other low median income census tracts. Wealthy neighborhoods tend to border neighborhoods that are nearly as wealthy, and lower income neighborhoods are more likely to border poor areas. In other words, there appears to be clustering based on median income. Several economic forces cause or explain the spatial grouping or clustering of income classes.

It's important to make sure that appearances are not deceiving—that the appearance of clustering is not an artifact of the way the map has been symbolized, for example. The degree of clustering will be measured statistically. The concept behind spatial autocorrelation is Tobler's First Law of Geography: "Everything is related to everything else, but near things are more related than distant things."[5]

In ArcGIS, a built-in tool, the (global) Moran's I tool (within the Spatial Statistics toolbox) allows users to determine, with a few clicks, whether a pattern in a given study area exhibits spatial autocorrelation. The Moran's I tool addresses the global question: Is there clustering (yes or no)?

Moran's I[6] uses feature values to identify and measure the strength of spatial patterns. Moran's I looks at the differences in values between each pair of neighbors to a given feature and then compares these difference to the differences in values between features in the overall study area. If the differences in values between neighboring features are less than the differences between all features in the study area, the values in that neighborhood are clustered.[7]

Moran's I takes on values between minus one and one. A value of zero indicates that the pattern is random. Similarly to the Average Nearest Neighbor, this is the "base case" or null hypothesis for the test of clustering based on Moran's I. If the value of Moran's I is statistically significantly closer to one, there is clustering. If the value of Moran's I is statistically significantly closer to minus one, there is greater dispersion than expected from a random distribution. To determine whether the calculated value of Moran's I is statistically significantly different from zero, the critical values of the z-score or p-value are used.[8] Without looking at statistical significance, there is no basis for knowing if the observed pattern is just one of many, many possible variants of a random distribution reflected in the sample.

The chapter previously noted that some economic forces can account for clustering. Economic and other forces can also account for dispersion (the opposite of clustering). Territorial animals (wolf packs) tend to be more dispersed than expected if wolf packs were distributed randomly. Wolf packs—and some retail outlets—carve out a territory and make efforts to keep competitors out (see chapter 4).

Statistical analysis supports the assumption that there is spatial autocorrelation for the Santa Clara County median income data. The output from the (global) Moran's I tool is given in figure 8-9.

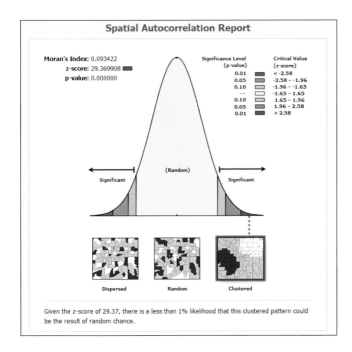

Figure 8-9. Spatial autocorrelation report for median household incomes by census tract in Santa Clara County.

Where the clusters are: Hot spot analysis

It has been determined that statistically significant clustering exists for the data on median household income by census tract for Santa Clara County. The next step is to determine the locations of clusters of high income and low income. One may ask, isn't this just like looking at a choropleth map of median income (see figure 8-3a)? Darker areas are census tracts with high income; lighter areas are census tracts with low income.

Not quite. A hot spot—in this case a cluster of high incomes—is not just a single census tract with high income but a group of census tracts of high income. A high income hot spot is defined, not just in terms of whether the income in a particular census tract is high, but whether it is in a neighborhood of census tracts which all have high income. What is meant by "high"? In hot spot analysis, high (and low) are measured in terms of *standard deviations from the mean*. There is one tricky part. Although a hot spot is defined in terms of the neighborhood, how is the neighborhood defined?

Hot spot analysis will be applied to the median household income data from Santa Clara County and to our data about California community college enrollment. The data about median incomes are associated with polygons. The data about community college enrollments are associated with points. The Hot Spot tool of ArcGIS can be applied equally well to both types of data.

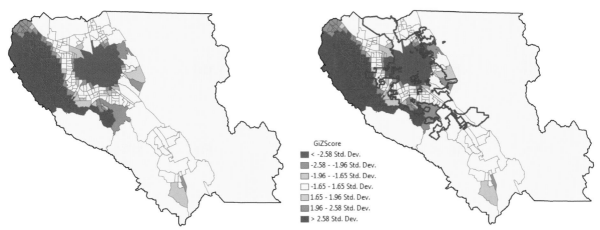

Figures 8-10a and 8-10b. Hot spots and cold spots of median household income by census tract in Santa Clara County at left, and superimposed with the boundary of San Jose (in purple) at right. At center is the map legend for the hot spot analysis. Courtesy of US Census.

GiZScore
- ■ < -2.58 Std. Dev.
- ■ -2.58 - -1.96 Std. Dev.
- ■ -1.96 - -1.65 Std. Dev.
- □ -1.65 - 1.65 Std. Dev.
- ■ 1.65 - 1.96 Std. Dev.
- ■ 1.96 - 2.58 Std. Dev.
- ■ > 2.58 Std. Dev.

Figures 8-10a and 8-10b show the results of applying the Hot Spot tool. High income clusters are identified in dark red; low income clusters are identified in dark blue. In figure 8-10b, the boundaries of San Jose, the largest of fifteen municipalities in Santa Clara County, are overlaid on the hot spot map.

Remember that the hot spots are statistically significantly high income areas close to other statistically significantly high income areas, and the cold spots are statistically significantly low income areas close to other statistically significantly low income areas.

Point hot spots can be analyzed similarly. There is a diversity of enrollment by campus among California public community colleges. But are there concentrations of enrollment? The global question (are there clusters?) and the local question (if so, where are the clusters?) can be addressed quickly using the tools in ArcGIS.

Applying the Moran's I tool to the enrollment data gives the answer to the global question. Figure 8-11 shows that there are clusters. Next, the Hot Spot tool is used to address the local question. Hot spots represent concentrations of campuses with high enrollment, and cold spots represent concentrations of campuses with low enrollment. The map showing hot spots is given in figure 8-12. The hot spots are in Southern California. (To those familiar with California development over the last hundred years, this result is not too surprising. Population throughout California has been rising over the last hundred years but has been rising faster in Southern California than in other parts of the state.)

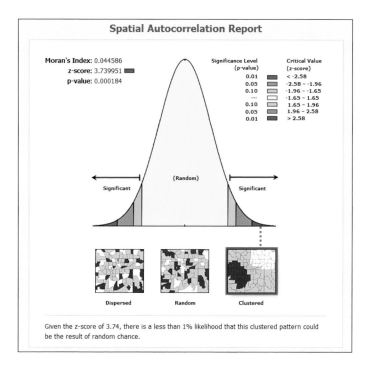

Figure 8-11. Spatial autocorrelation report for enrollment at California community colleges.

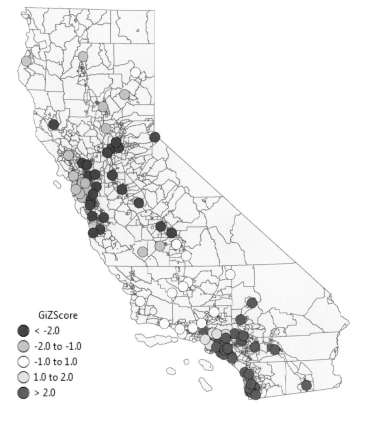

GiZScore
● < -2.0
◐ -2.0 to -1.0
○ -1.0 to 1.0
◔ 1.0 to 2.0
● > 2.0

Figure 8-12. Enrollment hot spots and cold spots for California community colleges, ranging in colors from red (hottest) to dark blue (coldest).
Courtesy of US Census; data and assistance provided by the California Community College GIS Collaborative and the California Community College System Office.

GIS-created datasets: Data generation unique to GIS

A GIS program can be used to create datasets that can be used within a GIS program or by other programs (like SPSS, SAS, Stata, or R) that would be difficult or impossible to create outside of the GIS program. This is because GIS has capabilities that mere database or statistical packages do not. Namely, it can connect datasets based only on location—an inherently geographic capability.

This can be done several ways, but one of the most direct methods is through the use of the **Spatial Join** technique, which requires analysis of the geographic relationships between the input datasets. (Database programs such as Microsoft Access can accomplish the equivalent of **Table Join**, which establishes a relationship between tables based on a shared field.)

Spatial join attaches the attributes of one layer to another layer and thus creates a new dataset. Spatial join may be used when the files do not have a common field, so the dataset that results from Spatial Join could not have been created by Table Join or any related method.

Spatial join is not a **commutative operation;** that is, spatially joining layer A to layer B is not the same as spatially joining layer B to layer A. This means one can do Spatial Join of two layers in two ways. Each way of doing the Spatial Join will result in a different dataset, and each way of doing Spatial Join will result in a dataset that could not have been produced by Table Join or its equivalent. Each of the different datasets produced by Spatial Join may be useful for economic development analysis—either within ArcGIS or by exporting it and using another statistical package for further analysis.

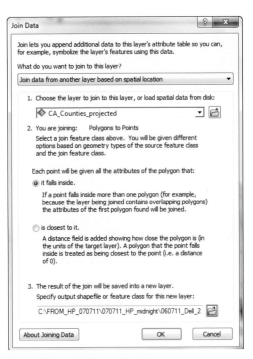

Figure 8S-2. Join Data dialog box

Using Spatial Join two ways: What to join with what?

Suppose there are two layers—a polygon layer of California counties (58 records) and a point layer of California community colleges (112 records) identified geographically by latitude and longitude. Because the community college file does not contain county identifiers, the two datasets cannot be joined based on a Table Join (figure 8S-2). The two sets can be joined, however, based on a Spatial Join.

Counties are joined to community colleges first. This operation creates a new (feature) dataset (a shapefile) with 112 records—the number of college campuses—and each address is now associated with the county in which that address resides both by the code number and the name of county (figure 8S-3). This information was not part of the original dataset. The other attributes of the county dataset were also "carried along" in this operation, and any additional county data can be added to the new dataset via the Table Join tool.

In the same way, Spatial Join can be used to attach any point shapefile or geodatabase features class (which may be an address dataset or just locations determined by using a GPS unit) the census tract or census block group. If census demographic or income variables were previously joined to the boundary file, those demographic and economic variables will be carried along to the newly created point shapefile.

continued ➡

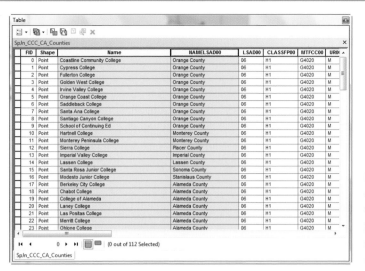

Figure 8S-3. Data (highlighted) created using Spatial Join one way. Data and assistance provided by the California Community College GIS Collaborative and the California Community College System Office.

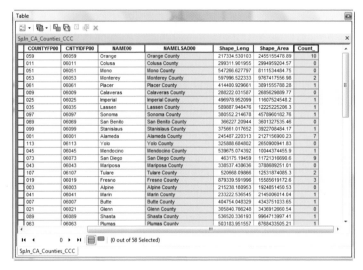

Figure 8S-4. County count data (highlighted) created using the Spatial Join tool the other way. Courtesy of US Census; data and assistance provided by the California Community College GIS Collaborative and the California Community College System Office.

In the example of the community colleges, the campuses can be spatially joined to counties. This will result in a dataset with 58 records—the number of counties in California—and an additional column of data concerning colleges. The "count," or number of college campuses in each county, will be found in the default case (figure 8S-4). These are new data that could not have been created by Table Join or any equivalent operation. The number of colleges may be a useful variable in the analysis, for example, as a factor that determines some other county-level variable, say, the labor force participation rate.

Similarly, the **Intersect tool** can be used to create datasets associated with point, line, and polygon files based on spatial location rather than common field. (For example, the multi-ring buffers of chapter 1 allowed the number of biotech firms at various distances from research universities to be counted.) Another tool, **Collect Events**, can be used with "incident data." One example of incident data is police reports. Often, the police are called to the same address more than one time. Each call may be recorded as a separate record in a database. The Collect Events tool takes the incident data and aggregates it based on location (and also displays the data using proportional symbols).

Regression analysis

An apocryphal saying states, "Correlation is not causation." Yet, economic development analysts and policy makers are keenly interested in causation. If they know how some variable x is causally related to some outcome y, and they can control or affect the value of x through policy actions, they can use the policy to affect the outcome y. Regression analysis is designed to model, estimate, and test such hypothetical causal relationships.

Independent and dependent variables

Economic models are abstract representations of the relationships among variables. For example, an economic model might suggest that variable y is determined by variables x_1, x_2, and x_3. This is represented generally by saying that y is a function of x_1, x_2, and x_3.

$$y = f(x_1, x_2, x_3)$$

For example, y might be job growth in an area, x_1 might be educational attainment in the area, x_2 might be distance to a community college, and x_3 might be an indicator (a "dummy variable") showing whether the area is or is not in a special economic zone. The term "dummy" is used in the sense of artificial, like a mannequin. Actually, dummy variables arise quite naturally in economic development analysis.

In the expression above, the variable y is a function of or depends on the variables x_1, x_2, and x_3. The variable y is called the dependent variable, and the variables x_1, x_2, and x_3 are called the independent variables. The variable y is the variable to be explained, and the x variables are the explanatory variables.

Starting with $y = f(x_1, x_2, x_3)$ as a general description of the relationship between y (the dependent variable) and the x's (the independent variables), a particular functional form can be adopted for the hypothesized relationship, for example, a linear form:

$$y = a_0 + a_1 x_1 + a_2 x_2 + a_3 x_3.$$

Regression is a method to estimate the **coefficients** (the a's in the equation above) of the hypothesized relationship, using data on the dependent (the y) and independent (the x's) variables. The estimated values of the a variables give the *marginal impact* of the corresponding independent variable upon the dependent variable. That is, the estimated value of a_1 measures the impact of an increase in x_1 on y holding constant x_2 and x_3. Similarly, the estimated value of a_2 measures the impact of an increase in x_2 on y holding constant x_1 and x_3. The idea of examining the effect of one variable on another, holding constant other variables, is expressed in the Latin phrase *ceteris paribus* ("other things equal" or "other things held constant").

In terms of our earlier example, where y might be job growth in an area, the coefficient a_1 would measure the impact of educational attainment on job growth, a_2 would measure the impact of distance to a community college on job growth, and a_3 would measure whether inclusion in a special zone had a significant impact on job growth.

The determination of the effect of one variable, say x_1 on another variable, y, while holding constant all other factors that can influence y, is central to *policy analysis*. If x_1 is a policy variable (for example, a tax rate or the degree of stringency of a zoning requirement) and y is an outcome that is important to stakeholders, say the cost of housing, then determining the marginal impact of x_1 on y (isolated from the influence of other factors such as x_2 and x_3 that are known to affect y) is the precise policy issue to be addressed. In terms of our example about job growth, the coefficient a_2 would indicate the impact of opening a satellite campus of the community college on job growth.

Because the data on y and the x's will not "fit" the linear model exactly, the equation that is being estimated will contain error, represented by ε, as in:

$$y = a_0 + a_1 x_1 + a_2 x_2 + a_3 x_3 + \varepsilon.$$

The error is reflected in the regression "residuals"—the differences between the values of the dependent variable, which are observed, from the values of the dependent variable, which are predicted by the regression. If the residuals exhibit spatial autocorrelation, the regression model may be misspecified, that is, it may be missing important explanatory variables. Statisticians Samprit Chatterjee and Ali S. Hadi discuss several tests for detecting auto-correlation in data and several methods for dealing with it in the context of regression analysis. "One of the standard assumptions in the regression model is that the error terms ... associated with the i-th and j-th observations, are uncorrelated. Correlation in the error terms suggests that there is additional information in the data that has not been exploited in the current model. The correlation is referred to as *autocorrelation*."[9]

Error in statistics, confidence levels, statistical significance, and critical values

The essential difference between a statistical approach to analysis and a nonstatistical approach is not that a statistical approach is *free of error,* but that the statistical approach formally *accounts for error.* Generally, any set of observations on some phenomenon, say income of households in a town, is based on a *sample* of households, not a complete *census* of all households. Samples are used instead of complete censuses for several reasons. One obvious reason is cost. Generally, it is less expensive to take fewer observations than more observations. Another often less-appreciated reason is that quality control is easier to maintain for a smaller sample than for a very large sample or a complete census.

The selected sample is only one of a very large number of samples that might have been selected. Imagine redoing the sample of households in the town from the ground up to determine average income. That would mean randomly selecting another set of households. Chances are that all or most of the households in the new sample would be different from the ones selected for the first sample. The average household income can be computed based on this new sample, and undoubtedly it will be different from the average computed for the first sample. But, if the methods used to select the samples do not have a systematic bias, the results will not be that different. If the unbiased sampling of household income was repeated over and over, and average household income was determined based on these samples, the averages would be distributed around the same central point—the *true* average household income.

Of course, in practice just one sample is taken. The degree of error can be estimated, based on knowing the distribution of the variable and having used an appropriate unbiased sampling technique. Judging whether a value determined by a sample is *statistically significantly* different from a given value depends on knowing the degree of error. This determination will be based on critical values of a statistic that depends on the nature of the statistical distribution being studied.

Various kinds of variables are included in regressions. Some variables are continuous valued. For example, educational attainment (measured in terms of years of schooling) is a continuous variable. Some variables are discrete valued. For example, number of bathrooms in a housing unit is a discrete number. Some variables take on only one of two values (zero or one). These are called "dummy" (or "artificial") variables. Dummy variables arise naturally in economic development analysis. Surveys often ask for a "yes/no" response—smoker or nonsmoker, income above $100,000 or not, and so forth. The example at the beginning of this section, in which y might be job growth in an area, introduced a dummy variable, whether the area was inside a special economic zone or not.

Economic development theory may guide us about expectations to have for the sign and significance of the estimated coefficients. Sometimes, regression analysis validates expectations based on theory; sometimes the regression results vary from what the theory suggests.

Regression analysis has several best practices. One of these is to make scatter plots of the dependent variable and each potential independent variable. Such scatter plots are suggestive of variables that might be included in the analysis. ArcGIS includes comprehensive guides to regression analysis. The ArcGIS 10 Help document, "Regression analysis basics," is especially recommended.[10]

An example of regression analysis would be to attempt to explain the clustering of biotechnology firms by some factors that economic theory suggests might be important. Among the factors discussed in chapter 1 were highly trained workers, the affordability of housing for less well-paid workers, the proximity to research universities, and the proximity to community colleges.

This can be formulated as an equation:

> *Number of neighboring firms*
> $= F(male\ doctorates\ in\ the\ census\ tract, female\ doctorates\ in\ the\ census\ tract,$
> *median gross rent in the census tract, distance to the nearest research university,*
> *and distance to the nearest community college).*

This should be viewed as a "first cut" or exploratory regression model. The explanatory variables are suggested by the theories discussed, though it may be valuable to include additional variables.

There is an **ordinary least squares (OLS) regression** tool within the ArcGIS Spatial Statistics toolbox. With just a few clicks, this tool allows users to estimate the regression coefficients associated with the explanatory variables, assess their individual significance, obtain diagnostic information about the overall quality of the regression, and get a visual representation of the *errors* or *residuals* of the estimated regression displayed on a map.

Running the OLS regression tool for the regression model mentioned above using the biotechnology data introduced in chapter 1 provides mixed results. The good news, briefly, is that the regression succeeds overall in explaining a significant part of the clustering of firms, and that some of the variables included have a strong influence. The bad news will come later.

The regression equation was:

> *Number of neighbors to a given firm*
> $= a_0 + a_1(Number\ of\ male\ doctorates)$
> $+ a_2(Number\ of\ female\ doctorates) + a_3(Median\ gross\ rent)$
> $+ a_4(Proximity\ to\ research\ university)$
> $+ a_5(Proximity\ to\ community\ college)$

The objective of regression analysis is to estimate the coefficients, that is, the a's. There are two aspects of this estimation: First, what is the value? Second, is the value statistically significant?

Table 8-1. Coefficient estimates and t-statistics for OLS regression

Coefficient	Variable	Estimated value	t-statistic	Statistically significant?
a_0	Constant term	24.777411	19.669862	Y
a_1	Number of male doctorates	-0.056055	-6.172506	Y
a_2	Number of female doctorates	0.047725	2.952642	Y
a_3	Median gross rent	0	0.966843	N
a_4	Proximity to research university	-0.000481	-11.148082	Y
a_5	Proximity to community college	0.000369	2.501687	Y

The two most important things about the entries in table 8-1 are *sign* and *significance.* (The *magnitude,* or the size of the coefficient estimates, is also of interest, but the first focus is on sign and significance.) All of the variables except median gross rent are significant; however, not all are of the expected sign. For example, the number of male doctorates in the census tract containing a firm has a negative or depressing effect on the number of its neighbor firms, while the number of female doctorates in the census tract has exactly the opposite effect—positive—and both coefficient estimates are statistically significant.

A rough rule of thumb is that a coefficient is statistically significant if the absolute value of the t-statistic is two or greater. The t-statistic will always have the same sign as the coefficient estimate—if the coefficient estimate is negative, the t-statistic will be negative. But the important thing is whether, disregarding the negative sign, the t-statistic is larger than two. The greater the t-statistic, the more statistically significant the coefficient estimate.

So, some of the good news is that one variable that has something to do with biotech firm clustering—distance to a research university—is of the expected sign (negative) and strongly statistically significant (a t-statistic of -11.148082). Proximity to community colleges, however, is another head scratcher. The coefficient estimate is positive—the greater the distance to the nearest community college, the lower the number of neighboring firms—and the coefficient estimate is statistically significant.

How good is this regression model overall? The ArcGIS OLS regression tool provides a wide array of diagnostic measures for regression analysis. Three are examined here.

The most important diagnostic statistics associated with regression are R^2, adjusted R^2, and the **t-statistics** associated with coefficient estimates. Briefly, these regression diagnostics can be interpreted as follows:

- R^2 is roughly the proportion of variation in the dependent variable (y) explained by the entire regression—all of the x's together
- Adjusted R^2 is R^2 minus a penalty for using too many independent variables. It is possible to increase R^2 by simply adding independent variables—"throwing in the kitchen sink"—making increasingly complex and theoretically poorly justified empirical models. The adjusted R^2 incorporates a penalty as a way of valuing parsimony in constructing empirical models
- The t-statistic is a value associated with each estimated coefficient. The value of the t-statistic shows whether that particular coefficient is statistically significantly different from zero. If a t-statistic indicates that a coefficient estimate is not statistically significantly different from zero, there can be no confidence that the associated variable has any impact on the dependent variable. As a rule of thumb, if the absolute value of a t-statistic is greater than or equal to two, the coefficient estimate is statistically significant.

These regression diagnostics tell us that the regression is a good start but not complete. The total amount of variation explained by the regression, the R^2, is 0.233476—23 percent. The Adjusted R^2 is not very different (0.226608). So, the regression is not using a lot of useless variables. In fact, it probably does not include enough variables. In addition to the standard output of regression packages, ArcGIS also provides residuals (errors of overestimation and underestimation) of the regression and maps the residuals (figure 8-13). It is important to map the residuals to understand whether the errors are clustered—whether they are spatially autocorrelated. If the residuals exhibit spatial autocorrelation, then very likely some variable associated with spatial distribution has been left out of the regression. In other words, the presence of *spatial autocorrelation in the residuals* of the regression is an indicator of a specification error. It is a signal to look for additional variables to explain clustering.

This map of residuals is symbolized like a hot spot map. Spatial autocorrelation is not desired in this case, but the presence of hot spots and cold spots means there is spatial autocorrelation, regardless. In addition, because the presence of spatial autocorrelation in the residuals (errors) of the OLS regression can indicate whether there is specification error, ArcGIS provides a straightforward means for testing using the global Moran's I to test for spatial autocorrelation of OLS regression residuals.

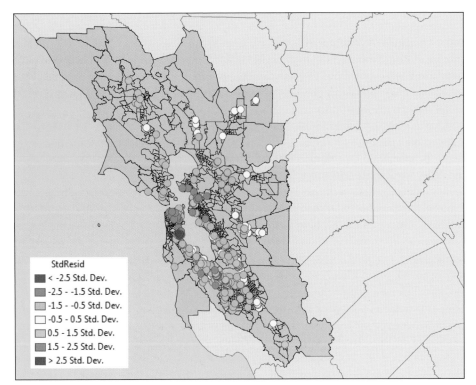

Figure 8-13. Map of the residuals for the OLS regression overlaid on San Francisco Bay Area census tracts. Courtesy of US Census; BayBio.

Hedonic regression

The term "hedonic" means pleasure (+) or pain (-). When used in connection with regression analysis, it is based on the view that consumers don't demand *commodities;* they demand *characteristics* that are embodied in commodities.

In a hedonic equation, the value of some commodity, say housing, is decomposed into the value of its characteristics. In this example the characteristics of housing include the floor area (F), lot size (L), number of bathrooms (B), and the quality rating of the public school district (S) in which the house is located. Then the house value, V, is decomposed into

$$V = a_0 + a_1F + a_2L + a_3B + a_4S + \epsilon.$$

The values a_1 through a_4 represent implicit prices. There is no explicit market, for example, in quality scores for public schools. But generally higher quality scores for local schools impart greater sales value to houses in the school's attendance zone. The implicit value of a higher rating for a school can be estimated using a hedonic regression.[11]

Elasticity and transforming variables

Many economic measures are expressed in terms of elasticities. **Elasticity** is a measure of how sensitive one variable is to changes in another variable expressed in terms of *percentage changes* (rather than *absolute changes*). There may be interest, for example, in knowing how much the quantity demanded of owner-occupied housing will change if there is a change in the price of owner-occupied housing. This can be cast in absolute terms, for example: "For every $10,000 increase in the price of owner-occupied housing, the quantity demanded of owner-occupied housing will decrease by 4,000 units." This sensitivity might be expressed as a ratio—the (absolute value of) change in quantity (4,000) by the change in price ($10,000).

That would give a ratio 0.4. However, this measure, based on absolute changes, has at least one problem. It depends on the units in which the variables are measured. Suppose housing prices were measured in thousands of dollars (000) instead of dollars. Then the ratio would be 0.0004.

If the *percentage* change in the price and the *percentage* change in the quantity are measured, the ratios do not change—the percentage changes are the same whether prices are measured in dollars or thousands of dollars or whether housing is measured in units or millions of units. Using elasticity makes it meaningful to compare the sensitivity of housing demand to price in Germany—computed in terms of Euros—with the sensitivity housing demand to price in the United States—computed in terms of dollars. Elasticity is a scale-free measure; it does not depend on the units in which the variables are measured.

Elasticity is easy to estimate in a regression. The idea of a linear regression was introduced with a function of the form

$$y = a_0 + a_1 x_1 + a_2 x_2 + a_3 x_3.$$

What if the hypothesis instead was that the relationship between y and the x's was nonlinear (and of the particular functional form given below)?

$$y = a_0 x_1^{a_1} x_2^{a_2} x_3^{a_3}.$$

The variables in this expression can be transformed by taking the natural logarithm (ln) of each side of the equation. Following the rules for working with logarithms, we obtain

$$\ln(y) = \ln(a_0) + a_1 \ln(x_1) + a_2 \ln(x_2) + a_3 \ln(x_3).$$

This expression is linear in the logarithms. Furthermore, each of the coefficients (the a's) are elasticities. For example, a_1 is the elasticity of y with respect to x_1; that is, a_1 is the percentage change in y for a 1 percent change in x_1. This estimate for the elasticity is generated automatically in a linear regression involving the transformed variables—it is just the coefficient estimate.

The most common elasticity is the own-price elasticity of demand—how demand for a good changes when there is a change in the good's own price. For example, the effect of changes in the price of owner-occupied housing price on the quantity demanded of owner-occupied units is represented by the own-price elasticity of demand for owner-occupied housing. Cross-price elasticities of demand are also estimated. The cross-price elasticity of demand is the effect on quantity demanded of one good when there are changes in the price of another good. For example, the percentage change in the quantity demanded of owner-occupied housing units can be estimated for a given percentage change in the price (rent) of rental housing units. There is an effect of rent on the demand for owner-occupied housing because owner-occupied housing units and rental housing units are substitutes.

Summary

Economic development analysis requires going beyond making pretty maps. A variety of indexes and statistical tools support economic development policy. ArcGIS brings many powerful statistical tools within easy reach in the Spatial Statistics toolbox. In particular, regression analysis—the analytic element at the heart of many policy issues—can be carried out in ArcGIS, with an appropriate account of the spatial context.

GIS can be used to construct meaningful indexes, and as you will see in the next chapter, raster data can also be used to construct indexes.

Notes

[1] Michael D. Ward and Kristian Skrede Gleditsch, *Spatial Regression Models* (New York: Sage Publishers, 2008): 339–49.

[2] Shown are the 87 campuses for which enrollment figures were readily available.

[3] See John F. McDonald and Daniel P. McMillen, *Urban Economics and Real Estate: Theory and Policy,* 2nd ed., (Hoboken, NJ: Wiley, 2011): 43.

[4] C. Gregory Knight, forward to *Spatial Autocorrelation: A Primer* by Daniel A. Griffith (Washington, DC: Association of American Geographers, 1987): iii.

[5] Waldo Tobler, "A Computer Movie Simulating Urban Growth in the Detroit Region." *Economic Geography* 46(2) (1970): 234–240.

[6] See Esri Help File "How Spatial Autocorrelation: Moran's I (Spatial Statistics) works."

[7] For further explanation, see Andy Mitchell, *The ESRI Guide to GIS Analysis,* Vol. 2, (Redlands: Esri Press, 2005).

[8] See ArcGIS 10 Help "What Is a Z-Score? What Is a P-Value?"

[9] See Samprit Chatterjee and Ali Hadi, *Regression Analysis by Example,* 4th ed., (Hoboken, NJ: Wiley, 2006): 197

[10] See also Luc Anselin, "Spatial Statistical Modeling in a GIS Environment" *GIS, Spatial Analysis, and Modeling,* eds. Maguire, Batty, and Goodchild (Redlands: Esri Press, 2005).

[11] See John F. McDonald and Daniel P. McMillen, *Urban Economics and Real Estate,* 2nd ed., (Hoboken, NJ:Wiley, 2011).

9

Working with raster data and imagery in economic development analysis

Objectives

- To illustrate the simple structure of raster data and demonstrate their uses for economic development projects
- To provide examples of common raster data, including aerial photography, elevation/terrain data, and raster data converted from vector map features
- To describe the ability of raster data to represent continuous geographic phenomenon, such as terrain and geocoded point clusters
- To demonstrate an economic development application using raster data by conducting a site-selection analysis for a new coffeehouse, using census data, land-use zoning, and ModelBuilder
- To demonstrate that raster data can be used to create specialty indexes—relative measures—that can incorporate weights of importance in a multivariate GIS analysis

This book has provided examples of geospatial data represented in ArcMap as points (such as customer address locations), lines (such as transportation networks), and polygons (such as jurisdictional boundaries). This method of representing geographic features in the vector data format is powerful and accurate although, as touched upon in chapter 3, there is a second method of representing geographic features, using a grid-based format known as raster data, in which geographic information is stored in an array of pixels, or cells. Each of these data formats has its advantages, and understanding both is an important component of project design so a project team can determine when it is appropriate to use either, or both, in an economic development analysis. This chapter explains the structure of raster data and their advantages for data display, standardizing analysis inputs, and ability to generate specialty indexes for site-suitability studies.

The simple structure of raster data

One way to visualize raster data is to consider the analogy of a photograph in a newspaper. While the photograph appears to be a "solid" image when viewed from a distance, it is actually composed of a tightly arranged matrix of dots of varying shades. In black and white photos, these dots are rendered in various "grayscale" shades, some darker and some lighter than others. Similarly, a digital raster file in GIS consists of a matrix of cells (pixels)

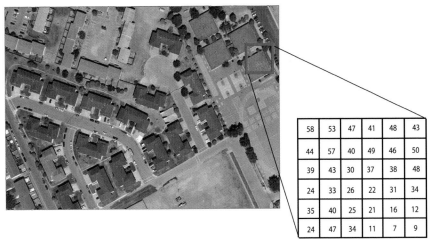

Figure 9-1. Aerial photographs are examples of information stored in the raster data format. Here, a portion of the image has been highlighted to reflect the grayscale values in each cell that together form the composite image. Each cell in the photograph is tagged with a value from 0 (white) to 255 (black), which informs GIS how to render the image on the screen. Courtesy of City and County of San Francisco Enterprise GIS.

arranged in a grid, in which each cell possesses a value, perhaps representing height above sea level, land-use type, or population density. In figure 9-1, a digital aerial photograph in GIS is shown, along with a portion highlighted to reflect the underlying raster cell values, with grayscale values ranging from white (0) to black (255).

In figure 9-2, the left image shows how simple features (polygons representing different land uses) would be represented as vector data in GIS; the right image shows how these same features might be represented using the raster format. In a GIS, both formats can be used simultaneously. Using ArcGIS Spatial Analyst, the GIS user can convert from one format to the other as project needs dictate. Such a conversion might be needed, for example, to design a project that requires all input datasets to be in the same data format for ease of comparison—two examples will be examined later in this chapter.

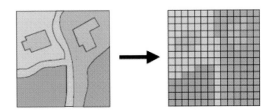

Figure 9-2. Showing vector features (left) representing buildings (red), land uses (yellow and green), water features (blue), and one possibility of how these features can be represented using the raster model (right). Each cell in the raster dataset would retain a numerical value corresponding to the original vector attributes (such as land-use type).

How are raster data useful?

Just like vector datasets, raster datasets can be previewed and organized into file geodatabases using ArcCatalog and then added to ArcMap's Table of Contents as map layers. One common example of this simultaneous use of vector and raster data is to add an aerial photograph (raster) to a map document and overlay it with vector datasets to assist with orientation, as shown in figure 9-3. This detailed aerial photograph of San Francisco using the imagery data can be freely downloaded from the city's website (or added to your map via an Internet connection, using the ArcGIS Online base map service). Overlaid upon the photograph is a polygon vector dataset (in yellow), representing census tract boundaries. Used together, these two different yet complementary datasets help us more easily interpret and navigate our project area.

Figure 9-3. Vector features (census tract boundaries of the San Francisco Bay Area, in yellow) and raster features (aerial photograph) displayed together in ArcMap. Courtesy of City and County of San Francisco Enterprise GIS.

So why is raster data more important when vector data seems more than adequate for representing geographic features in a GIS map? Rasters are highly useful for at least two primary reasons. First, they are a data format that levels (standardizes) the way that *all* geographic features can be represented in a map, including

- ground features, such as roads, buildings, and geocoded customer locations;
- quantitative data, such as demographic variables or labor participation rates;
- images, such as aerial photographs that can serve as useful backdrops to maps;
- surfaces, such as terrains, that are valuable for their ability to display landforms.

Additionally, the simple grid-based structure of raster data allows the display of all of this information to be standardized and overlaid grid-upon-grid and cell-upon-cell, thus opening the door to a rich set of mathematical and geoprocessing capabilities that are much less effectively accomplished with vector data alone. Multiple raster datasets can be overlaid to conduct multivariate analysis, using all of the cell values that reflect information for the same area of coverage.

Consider an example in which the relative measures of transit and parking accessibility in a community are examined, perhaps as part of a study interested in analyzing ease of access to a proposed transit-oriented development (TOD) site. Figure 9-4 illustrates three raster datasets for this purpose, all generated using ArcGIS Spatial Analyst, which contains geoprocessing tools specifically designed to work with raster data. The top layer in this diagram represents relative measures of transit accessibility based on proximity of residences to transit stations. Each cell in this dataset contains a value that represents the distance, in meters, from stations, calculated using ArcGIS Network Analyst. The second raster layer shows relative ease of access to parking lots, represented with varying shades of green, based on the same distance-generating process used for the transit stations. When each overlaying cell value from the first two layers are summed, the bottommost layer shows the result—a composite raster of access to *both* public transit and parking is produced, the darker orange areas representing those parts of the project area with the best access.

Such analysis can be performed with raster data (but not with vector data) because mathematical operations can be performed between overlaid raster cells. Furthermore, the relative importance of each input raster can be weighted; perhaps access to transit is assigned a weight of 75 percent importance and, because a potential TOD site

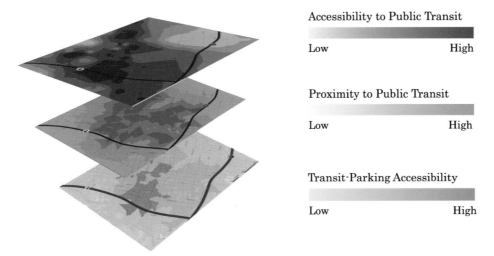

Figure 9-4. Multiple raster datasets can be overlaid and mathematical operations performed between multiple datasets to create a composite raster. Here, raster cell values measuring access to public transit (top layer) are added to cell values representing proximity to public parking (middle layer) to reveal a summed, composite layer that displays areas of a city that are the most convenient from both a transit and parking perspective. Image created by Diana Pancholi (May, 2010).

Figure 9-5. Through a GIS process known as georeferencing, digital raster images from non-GIS sources, such as scanned paper maps, can be brought into ArcMap and aligned with other datasets. Here, an 1860s map of downtown San Francisco (beige rectangular image) has been aligned with vector features representing today's streets, parks, and water bodies. Courtesy of City and County of San Francisco Enterprise GIS.

is being examined, parking is only weighted at 25 percent importance. An example later in this chapter will illustrate the weighting tools available in ArcGIS Spatial Analyst for site suitability studies.

Another interesting use of raster data in GIS is the ability to import scanned images (perhaps from paper maps, books, or reports) and, through a technique known as **georeferencing**, align these images to other datasets in your project map by identifying locations common to all the datasets. In figure 9-5, a raster image of downtown San Francisco from the 1860s was placed on a scanner, converted to a digital file, stored in a project geodatabase using ArcCatalog, added to ArcMap, and georeferenced by aligning it to street intersections that appear in both the historic map and in the GIS vector dataset representing streets. Giving "new life to old maps" provides a fascinating way to view urban change over time.

Raster data in economic development analysis

This is a good time to be involved with GIS technology; increasingly, it is becoming a matter of using just a few clicks to add rich and varied vector and raster data directly into ArcMap with minimal fuss and bother, either from a local file system or remotely via the web. Just a few years ago, GIS users could only dream of such functionality since, for example, adding current aerial imagery to maps used to be a cumbersome and expensive process.

The ability to combine both vector and raster data seamlessly in ArcGIS software allows myriad applications and workflows that will likely inform the work. For example, consider an economic development study focused on an undeveloped site on a community's suburban fringe. Vector data features could form the foundation of a project map and include the boundary of the study site, parcels, local roads, water features, and jurisdictional boundaries. This base map, as it is commonly known, can serve as the foundation for an overlaid aerial photograph obtained from a local vendor or from an online source. Another raster dataset might be added, representing the local terrain, from which steepness of slopes can be calculated for our study site, highlighting those areas that are suitable for urbanization (flatter) or conservation (steeper). When all vector and raster datasets are added to a project in ArcMap, the resulting map can be printed, and a storyboard can be created for a presentation or for posting on a website.

The adage that "a picture is worth a thousand words" is certainly the case with raster data in a GIS. Consider, for example, the many presentations that economic development officials need to make to elected officials. When describing development opportunities for a key site, the presentation invariably includes an aerial photograph of the site, supplemented by graphics and text to communicate clearly to the audience. If this infographic were created exclusively in a graphics software package, the result would be visually effective but would lack the analytical power that is embodied in a GIS since all digital spatial data in a GIS are linked to descriptive attributes. The power of imagery to convey a topic cannot be underestimated since a great deal of information can be conveyed effectively with images; it goes without saying that images are vital to an economic development professional when attempting to persuade and educate a target audience. Since all imagery is raster-based, it is important to know more about it in the context of a GIS.

Fortunately, aerial photography—likely to be the most common form of raster data used by economic development officials—is widely available for free or at a reasonable cost via a variety of web-based commercial and federal geospatial data portals (such as the US Geological Survey's National Map Seamless Server site) or from local and regional governments who have contracted with specialists to produce high-resolution aerial imagery for their jurisdictions. Additionally, Esri products such as the free ArcGIS Explorer application come equipped with up-to-date imagery. Also, imagery can be added directly into a map document via ArcGIS Online, which loads the images via the Internet from remote servers directly into an ArcMap document.

Raster data characteristics and categories

While the structure of raster data is simple, the format proves extremely powerful for sophisticated analyses. To begin with, a great variety can be introduced to the values attached to raster cells. Depending on the real-world feature being represented using the raster model, cell values typically fall into one of four classes:

1. *Category* (for example, values representing municipal zoning districts, such as residential, commercial, or industrial)
2. *Magnitude* (such as income in dollars or proximity in miles to a skilled labor pool)
3. *Height* (for example, the elevation of a location above sea level, in feet or meters)
4. *Spectral value* (values determined by the qualities of data collected by satellites—this can reveal variations in land cover, such as water, urbanized areas, forestry, and field crops)

Each of these classes can be depicted in raster datasets that fall into three general categories:

1. *Discrete data,* in which the features being classified are nominal or categorical in nature; that is, there is no natural ordering suggested by the information, and the real-world features represented consist of known and definable boundaries. Examples that might be used in an economic analysis include land use, census tract boundaries, or the zoning districts in a community that determine permitted land uses. Discrete data are also known as *thematic data* since a theme, or specific topic, of information is depicted.
2. *Continuous data,* in which the features represent phenomena that can be numerically ordered and that change continuously from one location to another with no clearly definable boundaries. Examples in an economic analysis might include the steepness of slope as a measure of a site's suitability for new development projects or varying proximities to transit services, based on the road network, as seen in figure 9-4. Temperature and elevation above sea level are other common forms of continuous data.
3. *Feature attribute data,* in which, for example, a photograph of a parcel of land or building might be stored in a geodatabase and linked to a geocoded point on a map representing the location of that property; clicking on the mapped location point reveals the raster image via a hyperlink. Tax assessors often use this technique for the management of tabular data linked with imagery of each assessed property in a jurisdiction.

A few words about raster data resolution

The term, **resolution,** often is encountered when working with raster data in GIS; fundamentally, the term refers to the size of the cells that comprise the raster dataset. Essentially, the smaller the size of the cells used to depict a geographic feature, the higher the level of detail and accuracy that can be achieved. Consider figure 9-6; note the

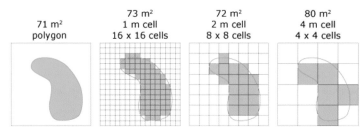

Figure 9-6. Original vector feature (left) and its representation in the raster model at increasingly "coarse" levels of resolution, reading left to right. Note the inverse relationship between raster cell size and the ability to mimic the original shape accurately.

polygon vector shape (it could represent a lake) at left becomes increasingly "degraded," going from left to right, while the lake is rendered with increasingly lower levels of resolution.

Note that as the number of cells decreases (in other words, lower resolution), the resulting raster dataset less accurately depicts the original polygon shape. The resolution of a raster dataset is defined by the equal length or height of each cell; in figure 9-6, note that the second, third, and fourth images from the left reflect a raster resolution in one meter, two meters, and four meters, respectively. Raster resolution and other attributes of any raster dataset can be reviewed in ArcCatalog, as seen in figures 9-7 and 9-8.

General	Source	Extent	Display	Symbology	Fields	Joins & Relates

Property	Value
☐ Raster Information	
Columns and Rows	410, 564
Number of Bands	1
Cellsize (X, Y)	50, 50
Uncompressed Size	451.64 KB
Format	ESRI GRID
Source Type	continuous
Pixel Type	unsigned integer
Pixel Depth	16 Bit

Figures 9-7 and 9-8. Attributes of a raster dataset—this one pertains to a terrain dataset in South Dakota—as viewed in ArcCatalog. Note the number of columns (410) and rows (564) that form this dataset, as well as the resolution—each cell in this dataset represents a portion of the earth's surface (fifty meters by fifty meters), as seen in the "Cellsize" information. This descriptive information is generated automatically by ArcGIS, based upon a scan of the file contents.

Understanding raster resolution has practical advantages for GIS projects. For example, if a vendor sends a digital aerial photograph with a resolution of one meter, the project team will know immediately that each cell in the photograph represents one square meter on the ground. If the photograph covers a large area, this is an excellent resolution to show ground features in great detail. With the development of increasingly sophisticated remote sensing equipment, it is now possible to use raster images with incredibly high resolution—some companies now provide aerial photographs with resolution measured in inches.

It might be easy to conclude that "bigger is always better" in terms of the number of cells in a raster file, but this is not necessarily so. While smaller cell sizes provide a crisper representation of ground features, this comes at the cost of large and potentially unwieldy file sizes that might be slower to display and render in a GIS. Determining an adequate cell size for a raster-based project is just as important in your GIS project planning stages as determining what datasets to obtain. ArcGIS Spatial Analyst offers the ability to resample a dataset's resolution, for example, to make the file size of a large image much smaller, though with the loss of some image crispness.

When considering the use of vector and raster data for a GIS project, it is helpful to weigh the advantages and drawbacks intrinsic to each data model. Vector data typically require less disk storage and are easy to maintain, and the points, lines, and polygons used in this model are better able to represent the actual shape of ground features. On the other hand, vector data utilize a more complex data structure than raster data and make map overlay analysis much more time-consuming than using raster data. Also, raster data require many multiples of disk space (depending on resolution) more than vector data, and the "blocky" appearance of raster data—evident when zooming in on a map to show fine levels of detail—might require the use of vector data for some map layers where the depiction of accurate boundaries, such as zoning districts—is critical to the analysis.

The main point to emphasize in this chapter, however, is that raster data are more than just pretty pictures. The chapter will now probe more deeply by considering examples of analytical possibilities using raster data.

Using raster data for terrain and point-cluster mapping

The simple structure of raster data opens the door to exciting uses for analysis and display of geospatial data. Raster data commonly are used to depict elevation or the terrain of a particular area, as seen in figure 9-8. These data are readily available from many vendors and via web-based download from sources like the National Map Seamless Server website, maintained by the US Geological Survey. In an elevation raster, also known a digital elevation model (DEM), each cell represents a specific portion of the earth's surface, and the elevation value (in feet or meters above sea level) at the center of that cell becomes the value that is affiliated with that cell. Once an elevation dataset is obtained and opened in ArcMap, it typically is rendered in grayscale values, as in figure 9-9, which represents a portion of the San Francisco Bay Area; the whiter areas represent higher elevations, and the black areas represent lower elevations. The hillier areas (concentrations of white cells) are easily distinguished from the stream valleys (linear concentrations of dark cells).

ArcGIS Spatial Analyst can be used to process these elevation data for analytical purposes. This extension to ArcGIS provides geoprocessing, analysis, and display tools geared specifically to working with raster data since input raster

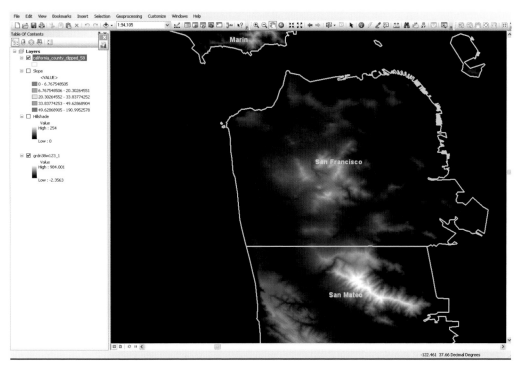

Figure 9-9. Preprocessed DEM data of the San Francisco area; lighter colors are cells representing higher elevation, while darker cells represent lower elevations. Vector data representing jurisdictional boundaries (yellow) are included to aid in orientation; the city and county of San Francisco comprises the majority of the map's geographic extent. The southern edge of Marin County lies to the north, and the northern portion of San Mateo County lies to the south.

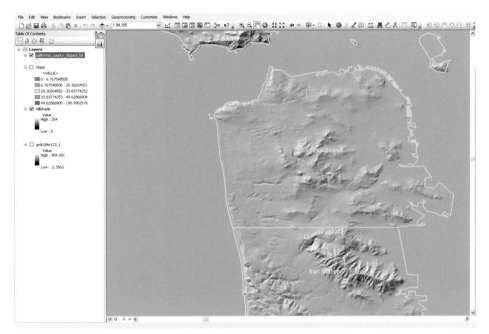

Figure 9-10. The DEM data after geoprocessing with the Hillshade tool in ArcGIS Spatial Analyst. This image might form the backdrop of an economic development project map, which needs to take steepness of slopes into account when locating a suitable property. As in figure 9-9, the city and county of San Francisco comprise the majority of the map's geographic extent, and the northern portion of San Mateo County lies to the south.

datasets can be manipulated to generate new datasets. One geoprocessing tool, Hillshade, applies a shadow effect to the preprocessed DEM data, using a user-defined sun angle and direction, and then produces a new raster dataset; the result is shown in figure 9-10.

It takes a few seconds to transform the original elevation data into a display that is eye-catching and represents the landforms in a more realistic manner, as seen in figure 9-10. The Slope geoprocessing tool from ArcGIS Spatial Analyst is applied next to create a new raster dataset from the original DEM data that shows the degree of slope within this portion of the Bay Area, as seen in figure 9-11. This analysis is particularly important to a site-selection process, which requires new construction to take place in areas that need the least amount of site grading.

Raster data are valuable when creating density maps for analyzing clustering and are particularly useful for locating relative densities of customers, businesses, or any other discretely located map features. Such maps reveal spatial patterns that may not be immediately evident by examining hundreds or thousands of individual points at once. Density maps are created using ArcGIS Spatial Analyst, using appropriate geoprocessing tools. Consider the example shown in figures 9-12 and 9-13.

Both images represent the same area—downtown San Jose (outlined in black). The image at left shows the location of all street intersections in the vicinity as vector points. ArcGIS Spatial Analyst can be used to effectively drape a grid of cells across these points (a raster resolution of twenty meters was chosen), and a value is added to each cell based on the count of intersections that falls within. The end result of this geoprocess, shown at the right, is a new raster dataset. Red depicts areas with fewer intersections, and green shows areas with more intersections. Such an analysis would be of great value to an economic development analyst seeking to locate an ideal site for a business that is heavily dependent on foot traffic, such as a downtown restaurant. More intersections generally equate to areas with higher urban density and pedestrian activity. In this scenario, the areas shown in green might be those that could be identified as a "first cut" when making important site-selection decisions.

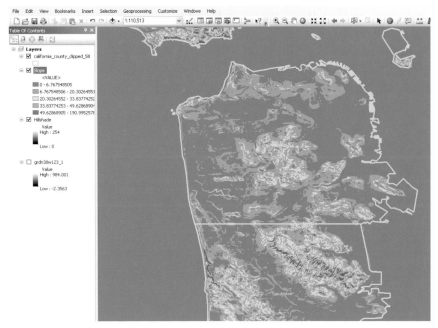

Figure 9-11. The DEM data after geoprocessing with the Slope tool. The default colors are shown, but they can easily be changed. Here, reds and oranges correspond to areas of steepest slope (note the mountain range toward the lower part of the image), and green areas represent least-steep (and absence of) slopes, including the bay and ocean that surround San Francisco.

Figures 9-12 and 9-13. Raster density maps show variations in clustering readily. The left image shows vector point features representing geocoded street intersections in downtown San Jose, California (the study area is represented with a black outline). After the points are geoprocessed using ArcGIS Spatial Analyst's density mapping tools, the right image shows variations in intersection density more easily—red areas are downtown locations with fewer intersections, and the green areas are those with more. Such a map might be used to help locate a restaurant site in locations with greater intersection density, reflecting more street activity. Images created by Paul Hierling.

Site selection using multiple, weighted raster datasets

This chapter has touched on site selection many times since it is a common practice of economic development professionals, and raster data are ideally suited to this purpose, as the next example will illustrate. This case study considers four inputs as part of the process one might undertake to find the best locations for a new coffeehouse in San Francisco: proximity to existing competitors, median income, household size, and city zoning regulations. The simple, grid-based structure of raster data allows for powerful geospatial analysis since many diverse data themes, such as the four listed here, can be standardized into grids using ArcGIS Spatial Analyst, and their cells' values can be overlaid and mathematically manipulated. Specifically, a suitability score will be assigned to each of the four inputs in terms of their likelihood of supporting a new coffeehouse; areas of the city with more desirable qualities, such as city zoning districts more supportive of a new coffeehouse, will receive higher scores. Conceptually, figure 9-14 illustrates what could be accomplished using the raster data scoring system.

The case study begins at the point when all of the input datasets have been collected. In this case, income and household-size data for San Francisco were obtained from the US Census Bureau, city zoning data were downloaded from the San Francisco Enterprise website, and geocoded competitor coffeehouse locations are drawn from earlier examples in chapter 7. All of these datasets have been evaluated for accuracy and stored in a project file geodatabase using ArcCatalog, according to the best practice procedures covered in chapter 3.

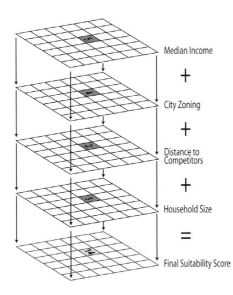

Median Income

+

City Zoning

+

Distance to Competitors

+

Household Size

=

Final Suitability Score

Figure 9-14. Conceptual diagram of the suitability analysis project. The suitability score values within each cell of the four input raster datasets will be summed to produce a final suitability score—cells with the highest score will represent parts of San Francisco most conducive to a new coffeehouse, based on proximity to competitors, income distribution, household size, and city zoning regulations.

The premise for this demonstration is that an economic development consultant wants to find suitable locations for a new, independent coffeehouse in San Francisco, which will appeal to young patrons and lie in close proximity to existing competitors. The consultant first created a detailed base map of the city in ArcMap, using the vector datasets (streets, parks, water bodies, and so on) that are freely downloadable from the city's website, as shown in figure 9-15.

Next, the consultant considers the four criteria that need to be met for the new coffeehouse location. Then a standardized scoring system is developed to weigh the relative values for each of these criteria. Let's examine each individually.

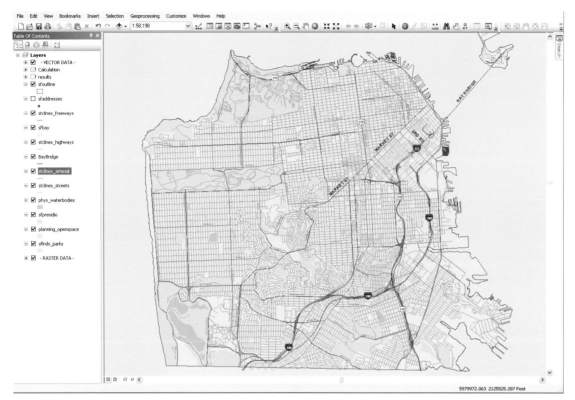

Figure 9-15. A base map of San Francisco created in ArcMap, using data available freely from the city's GIS website. Thematic datasets for the suitability project will be overlaid upon this base map in subsequent figures. Courtesy of City and County of San Francisco Enterprise GIS.

Criterion one: Areas of San Francisco zoned for commercial land uses

San Francisco has an unusually large and varied list of zoning districts to accommodate the multitude of land uses in its dense, diverse neighborhoods. A number of these districts permit the location of an independent coffeehouse "by right," meaning that no special conditions or variances are required from city officials. Such districts will receive the highest score in the suitability analysis, while other zoning districts will receive a lower score for other reasons. The suitability scoring chart is illustrated in table 9-1.

A map layer of these zoning districts can be added in ArcMap, and the suitability scores listed in table 9-1 can be added to the associated attribute table, using ArcMap's editing tools. Selected zoning districts tend to cluster in linear corridors, reflecting the presence of dense and walkable neighborhood shopping districts, for which San Francisco is known. Two of these districts can be seen prominently in figure 9-16. The Mission District commercial corridor lies north-south through the right-center of the city, and the busy Geary Street commercial corridor lies east-west toward the northwest portion of the city.

Table 9-1. Suitability scoring chart in relation to zoning districts

Suitability score	Zoning districts
5	Eight neighborhood center-focused zoning districts with a great deal of foot traffic that tend to attract small, independently owned businesses popular with young professionals.
4	Seventeen zoning districts that also accommodate neighborhood-centered businesses with high foot traffic but tend to be located along the periphery of neighborhood centers.
3	Zoning districts in downtown San Francisco's financial district. These areas are attractive because of the high demand for coffee throughout the workday, but they are assigned a middle score since the downtown area has a relatively low number of residences, raising concern that evening business activity would be low.
2	Zoning districts in the neighborhood where commercial land uses are supported, and which have a lot of foot traffic, and where there may be some support for a new coffeehouse but where tea has been more popular, according to your market studies. These areas will receive a relatively lower score as a result.
1	A score of one is assigned to residential zoning districts that allow for small, neighborhood-serving commercial development but which tend to be in much lower-density, auto-oriented areas with minimal foot traffic.

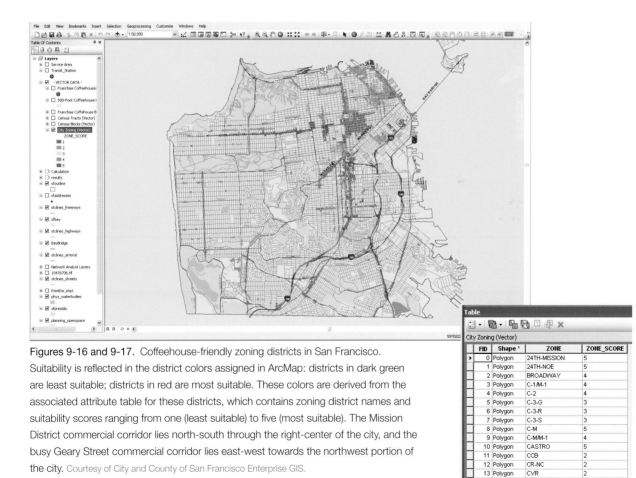

Figures 9-16 and 9-17. Coffeehouse-friendly zoning districts in San Francisco. Suitability is reflected in the district colors assigned in ArcMap: districts in dark green are least suitable; districts in red are most suitable. These colors are derived from the associated attribute table for these districts, which contains zoning district names and suitability scores ranging from one (least suitable) to five (most suitable). The Mission District commercial corridor lies north-south through the right-center of the city, and the busy Geary Street commercial corridor lies east-west towards the northwest portion of the city. Courtesy of City and County of San Francisco Enterprise GIS.

Criterion two: Locations within a quarter mile and a half mile of competing coffeehouses

The independent coffeehouse needs to be located in an area where a proven coffee-loving population already exists. Therefore, a suitable location will be within a half-mile (a walk of eight to ten minutes) from a competitor. Locations will receive an even higher score if they are within a quarter-mile (a walk of four to five minutes). The Buffer geoprocessing tool in ArcMap is used to create these distance areas, and then ArcGIS Spatial Analyst is used to convert them to a raster dataset. The resulting map layer is depicted in figure 9-18, and the associated attribute table is shown in figure 9-19. The suitability scoring chart is illustrated in table 9-2.

Table 9-2. Suitability scoring chart in relation to existing franchise coffeehouses

Suitability score	Description
5	Locations within one-quarter mile of an existing franchise coffeehouse
3	Locations within one-half mile of an existing franchise coffeehouse

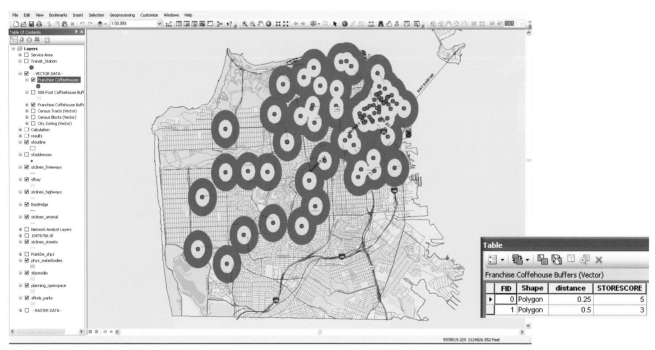

Figures 9-18 and 9-19. Multi-ring buffers showing proximity to existing coffeehouse competitors (in red). Also shown is the associated attribute table. Areas of the city that fall within the highly desirable quarter-mile buffer (yellow) receive a suitability score of five; areas that fall within the less-desirable half-mile buffer (brown) receive a score of three. As in the previous step, these buffers are converted to raster datasets using ArcGIS Spatial Analyst and will become part of a subsequent weighted overlay process. Courtesy of City and County of San Francisco Enterprise GIS.

Assigning suitability score values in this manner creates, in effect, a relative measure—a specialty index—that will be used later when weighting the relative importance of each of the four input datasets to determine a final suitability score.

Criterion three: Census tracts with a median age of less than thirty-six

The new store seeks to draw younger professionals because they tend to patronize coffeehouses, so the suitability analysis will favor census tracts with younger median ages, compared to citywide averages, as well as those age groups that tend to have higher levels of disposable income. These will receive higher scores in the analysis, as shown in the suitability scoring chart in table 9-3. Census data are freely available from the US Census Bureau's American FactFinder website.

Table 9-3. Suitability scoring chart in relation to median age

Suitability score	Description
5	Median age 33–35.9
4	Median age 30.1–32.9
3	Median age 27.2–30
2	Median age 24.3–27.1
1	Median age 21.4–24.2

As in the previous steps, the map layer is created, and suitability scores are assigned in the layer's attribute table. This third input dataset is then converted to a raster, as shown in figure 9-20.

Figure 9-20. A choropleth map displaying the median age by census tract in San Francisco. Colors represent coffeehouse suitability—darker colors are those tracts with a greater proportion of households with median ages that also signify a higher likelihood of households with children. These tracts are assigned a higher suitability score. Courtesy of City and County of San Francisco Enterprise GIS.

Criterion four: Census blocks with a median household size greater than two

Census blocks are the smallest unit of geography used by the US Census Bureau, and in a densely developed city such as San Francisco, census blocks typically correspond in size to actual city blocks. The assumption is that census blocks with higher household sizes reflect the presence of children, many of whom (teenagers, especially) will have some disposable income and will likely patronize the independent coffeehouse. Table 9-4 shows that the suitability analysis will give a higher score to census blocks with larger households.

Figure 9-21 shows the census blocks that "pass the test." In other words, blocks in darker colors represent those that are highly suitable for coffeehouse planning purposes; more lightly colored blocks are still suitable but less so than the darkest.

Table 9-4. Suitability scoring chart in relation to average household size

Suitability score	Values
5	Average household size > 4.59
4	Average household size 3.68–4.59
3	Average household size 3.05–3.67
2	Average household size 2.52–3.04
1	Average household size 2.01–2.51

Figure 9-21. A choropleth representing household size in San Francisco by census block. Colors represent coffeehouse suitability—darker colors are those blocks with a greater proportion of household sizes that signify a higher likelihood of households with children. These blocks are assigned a higher suitability score, based on the approach listed in table 9-4 and reflected in the associated attribute table affiliated with this choropleth map layer. Courtesy of City and County of San Francisco Enterprise GIS.

Now that the four inputs are finished, the consultant can overlay the four raster datasets in ArcMap and sum the cell values that are now "stacked." For example, a cell in an especially zoning-friendly district (five points) that also falls within an area with a high average household size (five points) and within a half-mile of an existing coffeehouse (three points) and within an area with our target median age (five points) would receive a total score of eighteen points. Other cells would receive higher or lower scores, depending on the interplay between the four inputs.

The consultant will total the scores using the Raster Calculator, which is available from the ArcGIS Spatial Analyst toolbox in ArcMap. When opened, the Raster Calculator allows for input raster cells to be mathematically combined, as represented conceptually in figure 9-22. The appropriate formula is entered into the Raster Calculator, as shown in 9-23; note the listing of available raster layers in the upper left, mathematical operators and functions toward the right, the entered expression towards the bottom, and the output raster file path at the bottom.

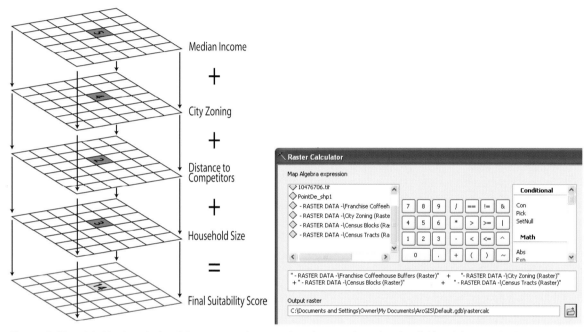

Figures 9-22 and 9-23. A reminder of the conceptual approach to the raster-based project (left) and the actual calculation using the Raster Calculator (right). The formula entered in the calculator allows for summing the values in the four input raster layers; the result will generate a new raster dataset with cells possessing the summed scores. Raster Calculator can be thought of as an index-generating tool to create specialty indexes for the analysis.

The result of the calculation appears in figure 9-25 as a new, summed raster layer in the map titled "Raster Calculator Results." The outputs of the raster calculator appear automatically as new layers in the Table of Contents. The symbology (colors) of this new layer has been assigned a light-to-dark color scheme to make it easy to find the most suitable locations for our new coffeehouse (that is, the raster cells with the highest total score).

The results of this examination show that portions of central San Francisco seem well-suited for a new coffeehouse from a zoning, competitive, and demographic perspective. Such an analysis might be used as a map-based component of an overall site-targeting strategy in addition to other marketing research. Although this example was designed to be simple and straightforward, it points to the ability of raster data to facilitate ease of computation between a wide variety of input variables and map features.

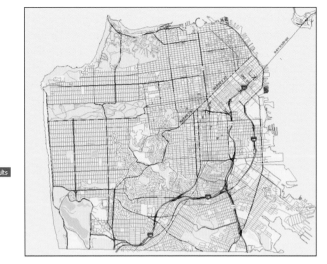

Figures 9-24 and 9-25. Raster Calculator results. Figure 9-24 at left shows the legend for the new map layer generated from the Raster Calculator; note that higher values correspond to cells in figure 9-25 at right, which totaled to a higher value after the four input rasters were summed. These areas are most supportive of a new coffeehouse, using the proposed approach. The raster cells that received the highest total score in the analysis are shown on the map in shades of red. According to the case study approach, these areas would be most suitable for a new coffeehouse in San Francisco. Courtesy of City and County of San Francisco Enterprise GIS.

Incorporating ModelBuilder into a weighted overlay site-selection process

An economic development professional who performs many site-selection analyses within the same geographic area would benefit from incorporating ModelBuilder into the work so that repeatedly used inputs and processes can be preserved for quick reuse. As seen in chapter 3, this component of ArcGIS lets users create a schematic diagram of linked input datasets, tools for processing operations, and output datasets, much like a flow chart—with no programming required. Preserving a geoprocessing analysis in a model allows for "what if?" scenarios to be performed quickly; for example, the existing coffeehouse buffer could be quickly adjusted to widths of one-third and two-thirds of a mile, based on research that identifies the willingness of customers to walk certain distances for their morning coffee.

ModelBuilder can be used to incorporate a specific geoprocessing tool—the Weighted Overlay tool—so that weights of importance can be assigned to different input variables. For example, assume now that city zoning policies ultimately are determined to be the most important site-selection criterion for a new coffeehouse and that demographic inputs, though important, are not quite as vital to the analysis. It is possible to assign a "percentage of importance" value to zoning that is higher than the one for demographic variables. This is what the Weighted Overlay raster tool does, as shown conceptually in figure 9-26.

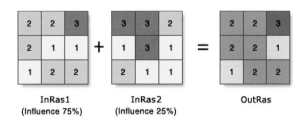

Figure 9-26. The concept behind the ModelBuilder Weighted Overlay tool. Input rasters are assigned a degree of importance totaling 100 percent; here, the first raster dataset has been assigned an importance weight of 75 percent and the second dataset a weight of 25 percent. The resulting output raster, shown at right, reflects the results of these two multiplication operations.

In this illustration, the two input rasters (in shades of brown and yellow) have been classified with a common measurement scale of one to three. Each raster is assigned a percentage influence (75 percent and 25 percent, respectively, as noted underneath each raster). Each cell value is multiplied by their percentage influence, and the results are added together to create the output raster (shown in green). For example, consider the top left cell in each step. The values for the two inputs become $(2 * 0.75) = 1.5$ and $(3 * 0.25) = 0.75$. The sum of 1.5 and 0.75 is 2.25. Because the output raster resulting from the Weighted Overlay geoprocessing tool is set to accept integer values, the sum is rounded to 2. The tool can be applied to the coffeehouse analysis by assigning weighted "importance" values as follows:

- For city zoning, 50 percent
- For proximity to competitors, 25 percent
- For census tracts with median age less than thirty-six, 13 percent
- For census blocks with a median household size greater than two, 12 percent

The total of these is 100 percent. This scheme places the greatest level of importance (50 percent) on the city of San Francisco's zoning regulations. Now the model can be created in ModelBuilder, as shown in figure 9-27 (your interface may vary slightly, depending on the version of ArcGIS software used).

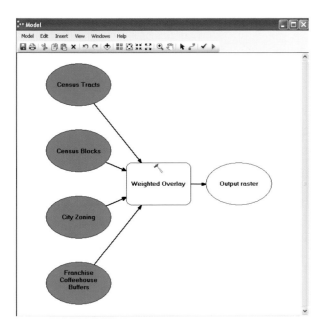

Figure 9-27. The ModelBuilder window with four analysis inputs added (blue ovals), as well as the Weighted Overlay geoprocessing tool and its eventual output raster. The tool and output are not colored in, indicating that the model is not yet in a ready-to-run state since the importance-weighting values must be entered into the Weighted Overlay tool.

The tool and output model elements are not colored in, indicating that they are not in a ready-to-run state and are awaiting decisions from the user. Double-clicking the Weighted Overlay tool in the model reveals its setup dialog. Here, weights of importance can be assigned to the four inputs, as shown in the "% Influence" column shown in figure 9-28.

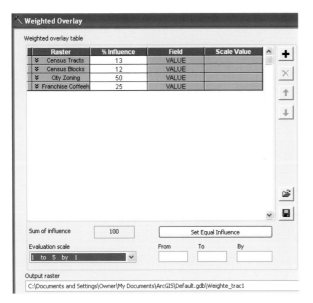

Figure 9-28. The Weighted Overlay dialog box. Note the listing of the four analysis inputs in the first column as well as their assigned degrees of importance in the "% Influence" column.

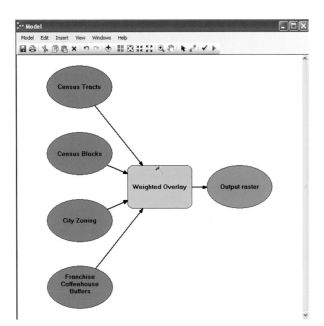

Figure 9-29. The weighted overlay model is now in a ready-to-run state since the importance settings shown in the previous figure have been entered into the Weighted Overlay tool.

Notice also that the "Evaluation scale" is set to "1 to 5 by 1" to match the scoring system used in our project (5 = most suitable for a coffeehouse; 1 = least suitable). When the Weighted Overlay tool is populated with these settings, the tool and the output element in the model should be colored yellow and green, respectively, indicating that the model has what it needs to execute.

Figure 9-30. The most suitable areas for a new coffeehouse (in red), as reflected with raster cells generated from the Weighted Overlay geoprocessing tool. Courtesy of City and County of San Francisco Enterprise GIS.

The model is now in a ready-to-run state. After the model executes, a new data layer will be added to ArcMap's Table of Contents. Zooming in on the map reveals the output of the weighted site-selection analysis, based on the four input raster sets, their respective scores, and the importance weights assigned in the Weighted Overlay tool. Red areas shown in figure 9-30 represent those that are most suitable for a new coffeehouse, according to updated project parameters.

Summary

This chapter examined raster data and the way their simple data structure allows for sophisticated geospatial analysis for a number of economic development applications. Aerial photographs can provide vivid map detail when used in conjunction with vector data for orientation. Multivariate analysis also can be conducted—and specialty indexes generated—by incorporating raster data into a GIS project. ModelBuilder, discussed in chapter 3 as a way to document and preserve geoprocessing workflows, was shown to be adept at managing multiple raster datasets and facilitating "what if?" analyses by letting users easily change inputs or settings, based on new or updated information.

Afterword

Looking forward, economic development increasingly will become more data driven and analytically heavy. This will coincide with increased competition for scarce funding and the need to show return on investment for activities, projects, and plans. For these reasons, GIS will be essential in economic development. Indeed, GIS is essential in economic development today.

Apart from these reasons, GIS is essential because it is a crucial component in the Geographic Approach—an analytic method that takes account of the essential spatial components of an economic development project that also will help guide future economic developers.

Two issues in particular likely will be at the center of future economic development: sustainability and regional interrelationships among agencies. Sustainability and green development are both important policy goals in many states and at the federal level. Insuring sustainable projects that comply with regulations will require detailed data and sophisticated analysis. We have argued in this book that GIS is the tool to solve that problem. Likewise, interregional cooperation will require an accounting of the spillovers—both positive and negative—of policies or projects in one jurisdiction upon its neighbors. Again, GIS is the natural tool for accounting for these effects.

GIS programs are becoming both increasingly sophisticated and easy to use. Web implementations of GIS are an especially hopeful development because they put an array of tools and data immediately in the hands of economic development officials. Cloud-based GIS also makes collaboration easier and increases the potential for communication with and participation of the public in economic development.

This book has examined a variety of economic development problems. It contrasted how economic development analysts addressed these problems before and after GIS became an essential methodology in economic development. The introduction mentioned the interplay of problems and tools. There is a tendency to put off addressing a problem if there are no tools to deal with it. We have described just a few of the hundreds of tools of GIS that address many problems faced by people involved in economic development analysis. Our goal has been to give a flavor of GIS in economic development analysis. If we have whetted your appetite, then we have met our goal.

Glossary

A

Add XY Data tool A tool in ArcGIS for plotting digital points based on latitude and longitude.

agglomeration economies Economies of scale (reduction in average cost) associated with the clustering of firms or people in a particular area; agglomeration economies include economies of localization (the decrease in average cost of firms in a particular industry associated with the concentration of firms in the same industry nearby) and economies of urbanization (the decrease in average cost associated with the concentration of people or firms or both, generally, not with reference to any specific industry).

apportionment problem The problem of associating data values from one geography (for example, police precincts) with another geography (for example, census tracts) when the geographic areas overlap but do not correspond exactly.

ArcCatalog A primary part of the ArcGIS suite, ArcCatalog is used to preview and manage geospatial data.

ArcMap A primary part of the ArcGIS suite, ArcMap is used to render maps and conduct spatial analysis.

ArcToolbox A primary part of the ArcGIS suite, ArcToolbox is a collection of hundreds of geoprocessing tools and functionalities.

attribute table Table of data that is directly linked to GIS map features. Each feature on the map (for example, a road or a customer location) has a corresponding set of attributes (for example, a road name or customer address) in the table. This linkage of map features and data tables enables both spatial and tabular analysis in a GIS.

Average Nearest Neighbor tool A tool in the ArcGIS Spatial Statistics toolbox that compares a given distribution of points (based solely on location, hence, not weighted) to a hypothetical random distribution. A statistical test determines whether the given distribution is statistically significantly more dispersed or more clustered than could be expected at random.

B

base map A map that incorporates basic data about an area; it depicts background reference information such as landforms, roads, landmarks, and political boundaries, onto which other thematic information is placed. A base map is used for locational reference and as the basis for making specialized maps for a particular analysis.

Business Analyst Desktop (see Esri Business Analyst Desktop.)

Business Analyst Online (see Esri Business Analyst Online.)

C

cannibalization The adverse effects on sales at one location by the presence of a retail outlet at another location.

census block groups The lowest level of geographic aggregation at which census data is readily available (see the sidebar, "Relationships among different geographies," in chapter 5).

central feature The feature from a collection of features that is the most centrally located. Identifies the most centrally located feature in a point, line, or polygon feature class.

centroid The center of gravity of a polygon shape.

choropleth map A map displaying variation in the value of a quantitative variable across a map area by using different colors or shades of the same color.

coefficients The unknown parameters or constants in a regression equation. The objective of regression analysis is to estimate the values of the coefficients. Regression diagnostics provide a statistical basis for assessing the significance of estimated coefficients. Coefficients measure the *marginal impact* of the corresponding independent variable upon the dependent variable.

Collect Events tool A tool within ArcGIS that creates a layer consisting of points at distinct locations, including a count of instances where multiple records occur at the same location.

commutative operation An algebraic operation (like addition) in which the order does not affect the outcome. For example, x + y = y + x.

D

database A digital record of information arranged as a series of linked tables consisting of fields (columns or variables) and records (rows or observations).

E

elasticity A measure of how sensitive one variable is to changes in another variable. Elasticity does not depend on the units in which the variables are measured because changes in the variables are expressed in percentage terms. For example, elasticity is used to express how sensitive supply or demand is to a change in price.

enterprise zone (EZ) A special zone in which some business developments receive advantageous tax treatment. EZ designations in California are awarded to local governments on a competitive basis.

Esri Business Analyst Desktop (BA Desktop) An extension of the main ArcGIS for Desktop software. It provides data and tools specific to business and economic development applications.

Esri Business Analyst Online (BAO) A browser-based GIS and business analysis program that combines data and tools for business and economic development analysis.

externalities Spillovers from the actions of one economic agent (a household or a firm) onto another economic agent. The externality can be beneficial (positive) or detrimental (negative). Examples include air pollution, traffic congestion, and lower crime rates because of increased police expenditures.

G

geocoding Creating digital points on a map based on address information.

geodatabase A database structure used primarily to store, query, and manipulate spatial data. Geodatabases store map features, tabular attributes of those features, and behavioral rules for data. Various types of geographic datasets can be collected within a geodatabase. Information is stored as a folder of files and can be used simultaneously by several users. ArcGIS supports two types of geodatabases: personal geodatabases and file geodatabases. Personal geodatabases are an older type based on Microsoft's Access database. File databases are preferred because they store more information.

geoprocessing A GIS operation used to manipulate GIS data. A typical geoprocessing operation takes an input dataset, performs an operation on that dataset, and returns the result of the operation as an output dataset. An example of a geoprocessing operation is buffering resulting in the depiction of rings representing quarter-mile distances from transit stations.

georeferencing The process of aligning geographic data to a known coordinate system so they can be viewed, queried, and analyzed with other geographic data. Georeferencing may involve shifting, rotating, scaling, skewing, and in some cases warping, rubber sheeting, or orthorectifying the data.

graduated symbols A method of displaying the variation in the value of a quantitative variable across a map area by using symbols of different sizes, for example, larger dots to represent bigger values and smaller dots to represent smaller values.

gross domestic product (GDP) A measure of aggregate output of an area (for example, a country or a state) over some period of time. GDP can be stated in nominal or real terms. Real GDP controls for inflation.

H

heat map A specific kind of choropleth map on which red is "hot," or high, and green is "cold," or low.

hot spot analysis A technique for identifying clusters of high values and low values in terms of statistical significance.

Huff Model An analytic model that can be used to evaluate market potential of a site. The model is implemented in BA Desktop and can be calibrated based on local data.

human capital The capital (productive capability) embodied in people in the form of knowledge and skills.

I

Intersect tool An operation in ArcGIS that identifies the common part of two data layers based on spatial location; similar to the intersection of two sets in mathematical set theory.

J

journey-to-work A term used to describe commuting between the place of work and the place of residence of the labor force.

L

labor productivity The value of output per worker (or per hour worked).

M

map layers The main geographic elements comprising a map document. Each map layer represents a particular feature of a geographic area, such as a street layer, a parcel layer, and so forth.

map packages Files with a .mpk file extension that make it easy to share complete map documents and their associated datasets with colleagues in a work group; all files needed to reproduce a map created by another author are packaged into one convenient, portable file.

market area The area in which a retail firm will be able to underprice its competitors (see applications in chapter 4 and chapter 6).

Mean Center tool Tool in the ArcGIS Spatial Statistics toolbox that determines the weighted or unweighted average location.

Measure tool Tool within ArcGIS that measures the distance between features.

Median Center tool Tool in the ArcGIS Spatial Statistics toolbox that determines the weighted or unweighted median location.

metadata Describes the content, quality, condition, origin, and other characteristics of data or other pieces of information.

Moran's I A statistic that measures the degree of spatial autocorrelation in data; ArcGIS has a tool to compute Moran's I in the ArcGIS Spatial Statistics toolbox.

N

Near tool Tool within ArcGIS to determine the shortest distance from one set of features (points, lines, or polygons) to another set of features (points, lines, or polygons).

O

ordinary least squares (OLS) regression The most commonly used technique of regression, which bases the estimated coefficient values on those values that minimize the sum of squared deviations of the proposed regression line from actual observations of the values of the dependent and independent variables. OLS is implemented as a tool in the ArcGIS Spatial Statistics toolbox.

P

procyclical Moves with or is correlated positively with the business cycle; for example, sales tax revenues are pro-cyclic—they increase in boom times and decrease during recessions.

Q

query (querying) The basic functionality of databases, including GIS databases; querying involves identifying characteristics of records in the database that the user is interested in; for example, all census tracts in a given area with a median household income greater than $50,000.

R

R^2 A diagnostic statistic in regression analysis that measures roughly the proportion of variation in the dependent variable (y) explained by the entire regression—all of the x's together; adjusted R^2 is R^2 minus a penalty for using too many independent variables. It is possible to increase R^2 by simply adding independent variables, in other words, "throwing in the kitchen sink," making increasingly complex and theoretically poorly justified empirical models. The adjusted R^2 incorporates a penalty as a way of valuing parsimony in constructing empirical models.

raster data Geographic data represented as pixels, including aerial and satellite imagery. As a data format, raster data represent the world as a surface divided into a regular grid of cells with numerical values (for example, elevation above sea level and land-use type) associated with each cell. Raster models are useful for representing geographic phenomena that lack distinct boundaries, such as landform, chemical concentrations, and air pollution.

regression analysis A method for estimating the significance of each factor in determining an outcome when several factors contribute independently to the outcome. Two methods of regression analysis, OLS and geographically weighted regression (GWR), are built into the ArcGIS Spatial Statistics toolbox.

resolution Refers to the detail with which a map depicts the location and shape of geographic features; the larger the map scale, the higher the possible resolution. As scale decreases, resolution diminishes and feature boundaries must be smoothed, simplified, or not shown at all; for example, small areas may have to be represented as points.

S

scale economies The phenomenon that, for some processes, the average cost of production declines as the scale (or volume) of production increases.

scatter diagram (scatter plot) A graph indicating the relationship between the values of one variable and the values of another variable.

Select by Attributes tool An ArcGIS tool that allows one to select elements of map features that correspond to specific user-determined conditions. An example is the selection of all census tracts that have a population greater than 5,000 residents.

shopping externalities The attractiveness to consumers of having a large and varied number of stores at one location, such as a shopping mall.

site selection Identifying a suitable location for a particular economic activity. Factors affecting the suitability include terrain characteristics, such as slope, soil, and proximity to transportation networks.

SmartMap search A capability within BAO and BA Desktop to search the entire Business Analyst database using user-specified characteristics.

Spatial Join An operation in ArcGIS that allows a connection to be made between datasets (layers) based solely on location.

Standard Distance tool Tool in the ArcGIS Spatial Statistics toolbox that determines a radius that is one, two, or three standard deviations from the weighted or unweighted mean center.

structured query language (SQL) The programming language of databases; it is used in ArcGIS, but a wizard, for example, in the Select by Attributes tool, automatically constructs the appropriate SQL statement for the user's query.

T

table (tabular) data Data organized in the form of a traditional database table, consisting of fields (titled columns) and records (table rows) (see database).

Table Join An operation in ArcGIS that allows a connection to be made between multiple datasets (layers) based on a common field or variable, similarly to other relational database programs like Microsoft Access.

tax expenditures The amount of taxes foregone in giving some special tax treatment to some entities, as with an enterprise zone; the concept of tax expenditure is defined with reference to a given tax regime.

thematic map Designed to convey information about a single topic or theme, such as population density or geology.

traces GPS tracks recorded by a GPS logger.

trade areas An area over which a given store will have an advantage over competitors.

transit-oriented development (TOD) Economic development that occurs around transit infrastructure, especially if the economic development is specifically planned around a transportation hub.

t-statistics Statistics associated with coefficient estimates in a regression that allow one to determine whether the estimated coefficient is statistically significantly different from zero—that is, whether the associated independent variable has a statistically significant marginal impact.

U

urban canyon effect The displacement of GPS recorded digital points due to the presence of tall buildings.

V

vector data Data format that represents geographic features using points, lines, and polygons. Vector models are useful for storing data that have distinct boundaries, such as country borders, land parcels, and streets.

vehicle-miles traveled (VMT) An aggregate measure of travel. The product of the number of vehicles multiplied by the number of miles traveled by each vehicle; one vehicle traveling two miles equals two vehicle-miles traveled.

W

workforce development Investment in training and education to meet industry needs, usually undertaken by community colleges and comprehensive universities.

Contributors

Randy Deshazo is a senior policy analyst at the Chicago Metropolitan Agency for Planning, working on freight mobility issues. Deshazo holds masters degrees in political science and urban planning from the University of New Orleans and University of Michigan, respectively.

Hongmian Gong is an associate professor in the Department of Geography at Hunter College of the City University of New York and in the Earth and Environmental Sciences Doctoral Program at the Graduate Center of the City University of New York. Her research interests are urban service economies in the United States and China and GIS/GPS applications in urban transportation.

John Lang is the chief economist for the city of San Jose and an economics lecturer at San Jose State University.

Index

best practices in economic development projects. *See also*
Geographic Approach: automating analysis with
geoprocessing, 61–63; describing methodology and
performing analysis, 53–55; documenting work
and sharing findings, 56–57; ensuring data quality,
57; identifying spatial data needs and creating
geodatabase, 49–50, 52, 194; managing and storing
data with file geodatabases, 60–61; presenting
results, 55–56; representing geographic features
digitally, 57–58; use of GPS technology in, 115–116;
using ModelBuilder for geoprocessing, 63–67

best practices in GIS, 43

best practices in regression analysis, 188

biotech industry clustering in San Francisco
Bay Area: housing prices and, 20; industry
challenges, 3; proximity to research universities,
13–17; public investment and, 12–13;
transportation infrastructure and, 17–20

boundaries: census tract, 51; jurisdictional, 6, 200; layers of,
13; overlaying on hot spot map, 182; ZIP codes and, 82

buffers, multi-ring, 16, 54, 61–62

Buffer tool (ArcGIS), 7, 16, 20, 61, 64–65

Business Analyst Desktop. *See* BA Desktop

Business Analyst Online. *See* BAO

calculated fields, 108–110

California community colleges: analyzing enrollment
statistically, 176–179, 181–183; drive-time areas,
36; geocoding locations of, 162–165; impact on job
growth, 186, 188-189; joining county and college
data layers, 184–185; workforce development, 12-13

California Enterprise Zone (EZ) program,
95, 97–98, 100, 102

cannibalization, 76

Case-Shiller Home Price Index, 107

causal relationships, 186

census tract data: boundaries, 51; census block groups,
73–74, 99; identifying with SQL statements, 103–
105, 109; previewing in ArcCatalog, 45–46;
reporting of, 99; for Santa Clara County,
California, 47, 96; specialty index of housing
costs, 108–109; statistical analysis of, 100–103

central feature, 176

Central Feature tool (ArcGIS Spatial Statistics), 176

central tendency, measures of, 173, 174–177

centroid, 174

choropleth maps, 20–21, 74, 175–176

clustering. *See also* biotech industry clustering in
San Francisco Bay Area: density patterns and,
171–172, 201–202; hot spot analysis, 172, 181–
183; identifying gas stations, 76–77; measuring
degree of, 180, 181–182; in regression analysis,
188–189; statistical significance of, 172

coefficients, 186–189

cold spot analysis, 182

Collect Events tool (ArcGIS), 185

colleges and universities in San Francisco Bay Area, 13–17

color ramp (map shading), 20, 102–103

community colleges, importance of, 12–13. *See
also* California community colleges

commutative operation, 184

commute times. *See* drive-time areas; journey-to-work data

competitors, identifying, 76–77

Consumer Price Index (CPI), 107

continuous geographic features, 58, 198. *See also* raster data

continuous variables, 187

correlation, 179

Council of Economic Advisors, 13

Create Address Locator tool (ModelBuilder), 166

Create Graph wizard (ArcGIS), 112

creating project geodatabase, 49–50, 52, 194

customer prospecting, 74–76

data. *See also* datasets: authors of, 47; quality of, 52, 57,
100; scalability, 67; summarizing, 54–55; vetting, 101

data formats, 120. *See also* raster data; vector data

Data Interoperability extension (ArcGIS), 119

datasets. *See also* data; site selection using raster datasets:
adding to ArcMap, 48; bundled with Business Analyst,
7; creating in GIS, 4, 40, 184–186; documenting,
52–53; with incomplete address information, 167;
multiple weighted, 203–205; organizing with
file geodatabases, 60–61; previewing, 45–46;
sources for compiling, 7, 21, 45, 52; streaming, 39;
Tapestry Segmentation, 25, 34–36; thematic, 39

degree of error, 186–187

demand, elasticity of, 191

density maps and patterns, 21, 171–172, 201–202

dependent variables, 186

descriptive statistics: definition of, 54; of variables,
100–102; viewing in ArcGIS, 100, 174

digital elevation model (DEM), 200, 201

digital pin maps. *See* geocoding

discrete geographic features, 57–58. *See also* vector data

discretionary trips, 123–124

dispersion and distribution, measures
of, 100, 173, 177–179, 181

documenting project results, 56–57

dot density maps, 21. *See also* geocoding

Dow Jones Industrial Average, 107

drive-time areas: around San Jose State University, 157–161;
determining with BAO, 35–36; inferring trip purpose
from, 123–124; sample reports generated by BA
Desktop, 129–132, 133–150; service areas, 168–170

Drive-time Trade Area tool (BA Desktop), 124

dummy variables, 186, 187

economic development. *See also* workforce development:
fundamental concepts in, 9–11; geocoding and,
156–157; GIS and, 4–6; sources of data, 8, 12, 21

economic development best practices. *See* best practices in
economic development projects; Geographic Approach

site selection for a retail store: applying the Huff Model, 78–82; customer prospecting, 74–76; defining the study area, 71–73; defining trade area around store, 74, 78, 82; forecasting sales, 80–82; Huff Model, 71, 72, 78–82; identifying competitors, 76–77; marketing strategy, 82; tax ramifications, 82–83
site selection using raster datasets: commercial zoning, 204–205; elevation data, 201–202; incorporating ModelBuilder with weighted overlay, 210–213; median household size, 208; proximity, 206–207; suitability scores, 203–209
SJSU. *See* San Jose State University drive-time areas
Slope geoprocessing tool (ArcGIS Spatial Analyst), 201–202
SmartMap Search, 26
social networks, 8
spatial analysis process list, 53–54
Spatial Join tool (ArcGIS), 7, 102, 184–185
specialty indexes, 108–111
spillovers, 6, 9, 11, 76–77
spreadsheets in economic development analysis, 4, 62–63
SQL (structured query language) statements, 103–105, 109
standard deviation, 173, 177–178, 181–183
Standard Deviational Ellipse tool (ArcGIS Spatial Statistics), 17, 177
Standard Distance tool (ArcGIS Spatial Statistics), 177–178
Standard Industry Classification (SIC) Codes, 9
Stanford University, 13–17
statistical approach vs. nonstatistical, 187
statistical measures: autocorrelation, 180–181, 189; average, 173, 174, 187; central feature, 176; correlation, 179; mean, 173, 174; mean center, 174–176; median, 173, 174, 175–176, 177; median center, 177; regression, 186–191; standard deviation, 173, 177–178, 181–183; t-statistics, 189
statistical significance, 172–173, 181–183, 187
Statistics of a Field function, 100, 111, 174
steps in Geographic Approach, 44–48
structured query language (SQL) statements, 103–105, 109
suitability scores, 203–209
Summarize feature (ArcGIS), 54–55
symbolization of maps, 102–103
Symbolizing quantitative data tool (ArcGIS), 7
symbols, graduated, 21

TAB files (MapInfo), 119
Table Join tool (ArcGIS), 7, 102–103, 184
tabular data, 7, 45–46, 102–103
Tampa Bay Area, Florida, 72–81. *See also* site selection for a retail store
Tapestry Segmentation database, 25, 34–36
Tapestry Segmentation report, 133–150
tax abatements, 82–83
tax incentives, 95
team leaders, 43
technological change, 8

technology transfer, 15–16
temporal autocorrelation, 180
terrain and point-cluster mapping, 200–202
thematic datasets, 39
thematic maps, 51–52, 72–73
time series analysis, 180
Tobler's First Law of Geography, 180
TOD (transit-oriented development), 9, 120, 168–170, 195–196
traces, GPS, 121, 126–128
trade areas, 74, 78, 82, 124
Trade Area wizard (BA Desktop), 124
trading entities, 9
traffic impacts in site selection, 83
transit-oriented development (TOD), 9, 120, 168–170, 195–196
transportation costs, 9
transportation infrastructure, 6, 17–20
travel. *See* drive-time areas; journey-to-work data
Travis County, Texas, 26–30, 41
t-statistics, 189

unemployment rates in San Jose, 96–97, 105
units of measure, 101
universities in San Francisco Bay Area, 13–17
unweighted average, 174
urban canyon effect, 120, 121, 126
urbanization, economies of, 6, 9, 11
US Census. *See* census tract data

variables: descriptive statistics of, 100–102; distribution of, 172, 173, 177–179; dummy, 186, 187; elasticity of, 191; independent and dependent, 186; modeling spatial relationships, 179–185; of relevance, 52
variance, 173
vector data: compared to raster, 59, 199–200; converting between vector and raster formats, 194, 206; representing discrete feature data, 57–58, 59, 194; types used in GIS, 7
vehicle-miles traveled (VMT), 116–117
vetting data, 101

wasteful commuting, 118
weighted average, 174
weighted importance values, 210–211
Weighted Overlay tool (ModelBuilder), 210–212
"what if" scenarios: fiscal and demographic, 108; in ModelBuilder, 67, 210; and SQL statements, 103–105
workforce development. *See also* biotech industry clustering in San Francisco Bay Area: definition of, 3; data sources for, 7, 8, 12–13, 21; effectiveness of, 21; policies, 12–13

ZIP Codes, 12, 15, 25, 31, 35–36, 82, 155, 167
zoning, 5–6, 117